T0314968

An Introduction to Cochran–Mantel–Haenszel Testing and Nonparametric ANOVA

An Introduction to Cochran–Mantel–Haenszel Testing and Nonparametric ANOVA

J.C.W. Rayner and G. C. Livingston Jr.
University of Newcastle
Australia

This edition first published 2023
© 2023 John Wiley & Sons Ltd

All rights reserved. No part of this publication may be reproduced, stored in a retrieval system, or transmitted, in any form or by any means, electronic, mechanical, photocopying, recording or otherwise, except as permitted by law. Advice on how to obtain permission to reuse material from this title is available at http://www.wiley.com/go/permissions.

The right of J.C.W. Rayner and G. C. Livingston Jr. to be identified as the authors of this work has been asserted in accordance with law.

Registered Office
John Wiley & Sons Ltd, The Atrium, Southern Gate, Chichester, West Sussex, PO19 8SQ, UK

Editorial Office
The Atrium, Southern Gate, Chichester, West Sussex, PO19 8SQ, UK

For details of our global editorial offices, customer services, and more information about Wiley products visit us at www.wiley.com.

Wiley also publishes its books in a variety of electronic formats and by print-on-demand. Some content that appears in standard print versions of this book may not be available in other formats.

Trademarks: Wiley and the Wiley logo are trademarks or registered trademarks of John Wiley & Sons, Inc. and/or its affiliates in the United States and other countries and may not be used without written permission. All other trademarks are the property of their respective owners. JohnWiley & Sons, Inc. is not associated with any product or vendor mentioned in this book.

Limit of Liability/Disclaimer of Warranty
While the publisher and authors have used their best efforts in preparing this work, they make no representations or warranties with respect to the accuracy or completeness of the contents of this work and specifically disclaim all warranties, including without limitation any implied warranties of merchantability or fitness for a particular purpose. No warranty may be created or extended by sales representatives, written sales materials or promotional statements for this work. This work is sold with the understanding that the publisher is not engaged in rendering professional services. The advice and strategies contained herein may not be suitable for your situation. You should consult with a specialist where appropriate. The fact that an organization, website, or product is referred to in this work as a citation and/or potential source of further information does not mean that the publisher and authors endorse the information or services the organization, website, or product may provide or recommendations it may make. Further, readers should be aware that websites listed in this work may have changed or disappeared between when this work was written and when it is read. Neither the publisher nor authors shall be liable for any loss of profit or any other commercial damages, including but not limited to special, incidental, consequential, or other damages.

Library of Congress Cataloging-in-Publication Data is Applied for:

Hardback ISBN: 9781119831983

Cover Design: Wiley
Cover Images: © metamorworks/Shutterstock, KatieDobies/Getty Images

Set in 9.5/12.5pt STIXTwoText by Straive, Chennai, India
Printed and bound by CPI Group (UK) Ltd, Croydon, CR0 4YY

C9781119831983_060223

John: For John Best, my friend and colleague for over half a century
Glen: For my boys, Huxley and Duke

Contents

Preface

My, that is, John Rayner's first acquaintance with the Cochran-Mantel-Haenszel (CMH) tests was through my long-term colleague and friend, John Best. We were undergraduate students together. Afterwards he made a career in the CSIRO, a government science organization in Australia, while I worked in academia. When we started researching together, we found that his focus was primarily on the applications and computational matters, whereas mine gravitated to the theory and writing. There was considerable overlap, so we could have meaningful dialogue, but there was sufficient difference such that we each brought something different to the collaboration. I like to think that together we achieved more than what would have been possible separately.

Our early focus was on goodness of fit testing, but of course we delved into other matters. For example, from time to time we tried and failed to find satisfactory nonparametric tests for the Latin squares design.

When I did come to grips with the CMH methodology, I was largely frustrated. Inference is conditional, and that is a dated paradigm. Software is not commonly available, so application is inconvenient. In addition data entry is often inconvenient. The design, that I call the CMH design, is very restricted. If I want a nonparametric test for categorical response data, I want it for a wide range of designs – such as the Latin square design. In the text of this manuscript we quite deliberately try to get as much as possible from CMH methods for the Latin square and balanced incomplete block designs. The outcome is far from satisfactory.

The nonparametric ANOVA methodology came about as an offshoot of partitioning the Pearson statistic. Once developed, it is natural to compare it with CMH, but there are issues here. The CMH tests are of two types: nominal and ordinal. To my mind the Pearson tests should be compared with the nominal CMH tests, and a natural competitor for the ordinal CMH tests is the nonparametric ANOVA tests. But the latter are able to assess differences in treatment effects beyond location, and if treatments are ordinal as well as the response, beyond the simple correlation. It was therefore necessary to develop the ordinal CMH tests so that they could assess higher moment effects, to enable a meaningful comparison.

The text following attempts to clearly and fully develop the CMH methodology with some recent enhancements. It is interesting that the Kruskal–Wallis and Friedman tests are particular cases of CMH mean scores tests, although they were, of course, developed from different perspectives. The completely randomised design and the randomised block design both get individual chapters in the text firstly because of their fundamental importance. Moreover the ordinal CMH tests in these cases have simple forms, and therefore are accessible to users of statistics who will benefit from applying them for these designs, and who will benefit from knowing these methods are available in more complex designs.

Towards the end of the text, we outline the nonparametric ANOVA methodology, so that the reader can make comparisons. In the first two papers developing these methods, the context was simple factorial models subsequently extended to higher order models. For this text we develop the methods for the CMH design, and then extend them. The presentation is, we believe, clearer than in the original papers.

We are aware there are issues with some of the tests derived using the rank transform method. Nevertheless these tests, and others, arise from the nonparametric ANOVA. It is worth noting here that nonparametric ANOVA starts with a table of counts of categorical response data. The Pearson test for association can be applied to such tables and decomposed into components. This is an arithmetic decomposition and requires no model. The components are indexed by what might be called 'degree', and it is natural to enquire what degree means and what the components tell us about the data. One way of investigating this is to apply the components of a particular degree to the ANOVA to which the data was to be applied. It is then of interest to investigate the properties and performance of these tests. Here we apply both the nominal CMH and nonparametric ANOVA tests to data for which both are applicable. In the cases where we have done so, we find similar outcomes.

Another important aspect of nonparametric ANOVA is that if we assume weak multinomial models, then inference based on components of different degrees are uncorrelated. Thus, for example, first degree inference does not influence second degree inference, and so on. In other words, a significant mean effect cannot cause a significant second degree effect.

The climax of this manuscript is the examples in the concluding chapter where we apply all the CMH and nonparametric ANOVA. Our conclusion from doing this is that although the tools are different, the conclusions are largely the same. A significant advantage of the nonparametric ANOVA is that it is more generally applicable than the ordinal CMH. For this reason we would recommend the use the Pearson tests instead of the nominal CMH tests, and the unordered and ordered nonparametric ANOVA tests instead of the ordinal CMH tests. However, the CMH tests are familiar to many users and appropriate in the areas of application of greatest relevance to them; their use is embedded. For these users we hope that the extensions and new results here enrich their future data analysis.

Another issue with the application of the CMH methodology is the availability of suitable software. For example, there does not appear to be an **R** function that is able to apply all four tests to a data set that allows for it. Other popular software such as SAS and SPSS do not easily allow for the application of all four tests. The manuscript and **R** package written in conjunction with it will hopefully provide a suite of examples and data sets to the reader. These will hopefully facilitate familiarity of the methodologies outlined for data analysts, such that they can be applied to their own data sets of interest. The R package is available at: https://cran.r-project.org/web/packages/CMHNPA/index.html.

J.C.W. Rayner & G. C. Livingston Jr

1

Introduction

1.1 What Are the CMH and NP ANOVA Tests?

The Cochran–Mantel–Haenszel (CMH) tests are a suite of four nonparametric (NP) tests used to test the null hypothesis of no association against various alternatives. They are applicable when several treatments are applied on several independent strata, or blocks. Every treatment is applied on every stratum. The responses are categorical, and so are recorded as counts. Two of the tests require scores, and so are called ordinal tests; the two that don't are called nominal tests. Ranks, or if observations are not distinct then mid-ranks, are often used as category scores. However, there are many other options, such as the class midpoints if the data are real-valued and continuous. Often 'natural' scores are convenient: just score the categories 1, 2, ...

From time to time in the material following we will refer to the CMH design, meaning that categorical responses are recorded for a number of treatments applied to a number of independent strata or blocks. The simplest CMH designs are arguably the completely randomised design (CRD) and the randomised block design (RBD). The methods do not directly apply to the balanced incomplete block design (BIBD) and the Latin square design (LSD). Both of these have, in a sense, missing observations.

One reason why the CMH tests are important is they provide a third option for the two simplest designs: the CRD and the RBD. For the CRD if the data are consistent with certain assumptions the parametric one-way ANOVA F test is available. If these assumptions aren't satisfied then the Kruskal–Wallis rank test is often applicable. If the responses are categorical then the CMH tests may be used. Similarly for the RBD the options of most interest are the parametric two-way ANOVA F test, the Friedman rank test, and the CMH tests.

An Introduction to Cochran–Mantel–Haenszel Testing and Nonparametric ANOVA,
First Edition. J.C.W. Rayner and G. C. Livingston Jr.
© 2023 John Wiley & Sons Ltd. Published 2023 by John Wiley & Sons Ltd.

Where possible we seek to give analyses that will prove to be as accessible as possible. The nominal tests are simply related to what are often called chi-squared tests, but we prefer to call them Pearson tests. These are well-understood and often available in 'click-and-point packages' such as JMP, and are on call in ®[1]. The Pearson tests are natural alternatives to the nominal CMH tests. We give a simple expression for the CMH correlation test. This expression involves familiar sums of squares and gives useful additional information as a corollary. For the CRD and the RBD we give simple expressions for the other CMH ordinal test, the CMH mean scores test.

This simplicity means that some analyses that can be done by hand, meaning pencil and paper. Some are better suited to 'click-and-point' packages such as JMP. Otherwise analyses are best done by computer packages.

In summary the CMH tests are nonparametric tests for categorical response data. The applicable designs include, but are not limited to, the CRD and the RBD. These designs are of fundamental importance in many areas of application.

The nonparametric (NP) ANOVA tests are competitor tests for the ordinal CMH tests. They apply to data sets for which a fixed effects ANOVA is an appropriate analysis. These tests involve transforming the responses, and possibly, if treatments are ordered, the ordered treatment scores as well, both using orthonormal polynomials. The ANOVA is then applied to the transformed data of given degree. The resulting analysis permits testing univariate treatment effects and bivariate treatment effects. Under weak assumptions tests of different degrees do not affect each other. The NP ANOVA methodology is available more generally than the CMH methodology.

1.2 Outline

A fundamental aim of this book is to introduce users of statistics to new methods in CMH and related testing. Before discussing the new CMH methods, it is necessary to give the old, or basic methods. This will be done in Chapter 2.

This is followed by a discussion of the CMH tests in the CRD and RBD. This will introduce the Kruskal–Wallis and Friedman tests that can be shown to be CMH tests. Although CMH methods are not directly applicable

---

1 Source: The R Foundation.

to the BIBD and the LSD, consideration will subsequently also be given to these designs.

Data for the traditional CMH tests are given as tables of counts. Inference assumes that all marginal totals in such a table is known before the data are collected. This is *conditional inference*, inference conditional on these quantities being known. There are dual methods for which the marginals are not known. This is *unconditional inference*. Confusion can arise from a lack of clarity as to whether a particular test is conditional or not.

Next we turn to what we call nonparametric ANOVA. The data may be unranked or ranked, categorical or not. The primary objective is to analyse the data using an ANOVA model available via the linear model platform inherent in, for example, JMP and ℝ. If only the responses are ordered the method enables higher moment effects to be scrutinised. If both treatments and responses are ordered, then as well as the usual order (1, 1) correlation, those of degree (1, 2) and (1, 2) may be assessed. These reflect umbrella effects. Higher degree generalised correlations may also be scrutinised.

Given that nonparametric ANOVA can assess higher degree effects, it is natural to generalise the ordered CMH tests so that they can do so too. It is then possible to give a comparison of analyses by both methods.

Our discussion will involve some well-known rank tests: the Kruskal–Wallis, Friedman, and Durbin tests. It is therefore sensible to make some general comments about ranking. Of particular interest is the treatment of ties. For general ranking methods see, for example, https://en.wikipedia.org/wiki/Ranking. For a treatment of ties in the sign test, see Rayner and Best (1999).

We now say what we mean by ranks. Given a set of data, $x_1, x_2, \ldots, x_n$, there exists a transformation t such that $\{t(x_i)\}$ are ordered. To say that $t(x_i)$ has *rank j*, namely, r_j, means $t(x_i) = r_j$ and $1 = r_1 \leq r_2 \leq \cdots \leq r_n \leq n$ (or the reverse set of inequalities). That the ranks are distinct means these inequalities are strict: $r_1 < r_2 < \cdots < r_n$. Ties occur when the ranks are not distinct. Of the many ways of dealing with ties, mid-ranks is perhaps the most used. Mid-ranks assigns to a group of tied data the mean of the ranks they would otherwise have been assigned. Thus if $r_{i+1} = \cdots = r_{i+j}$ then the mid-rank for these data is $(r_{i+1} + \cdots + r_{i+j})/j$.

Essentially ranking takes a set of n observations, categorises them into $c \leq n$ categories, and assigns to these categories distinct scores, b_h, say, with $b_1 < b_2 < \cdots < b_c$. Clearly for untied data $c = n$ and $b_h = h$ for $h = 1, 2, \ldots, n$. Suppose that if an observation of treatment i falls in category h then the indicator variable $N_{ih} = 1$, and zero otherwise. Then $b_h N_{ih} = r_{ih}$, the rank for this observation.

1.3 ®

We have written an ® package called CMHNPA, which will serve as an accompaniment to this text. The package contains all the data sets which are analysed as well as functions written for the statistical methods and techniques discussed. Within each of the chapters there is ® code where example data sets are used. If the output from the functions is excessive, it will sometimes be suppressed; however, the code will be presented for the reader to execute the functions themselves.

In the following set of example code, the package is loaded along with the car package. Code is also shown to attach a data set called dataset to the workspace. Attaching the data files to the workspace allows the variables within the data frame to be accessed directly when using the functions, as opposed to using dataframe$variable syntax.

```
library(CMHNPA)
library(car)
attach(dataset)
```

All of the code that follows in this text has the type of code shown above omitted. Therefore, if the reader wishes to recreate the output in later chapters, the packages will need to be loaded, and the data set attached to the workspace.

The ® package is currently available from: https://cran.r-project.org/web/packages/CMHNPA/index.html. It will undergo ongoing development and so output of functions may change and additional options added for functions over time.

This brief introduction will be ended with two examples. The first involves a non-standard design to which the CMH methods do not apply. The purpose is to demonstrate another nonparametric approach: the rank transform (RT) method discussed by Conover and Iman (1981). If, in performing a parametric test, the assumptions are found to be dubious, then the idea is to replace the original data by their ranks – usually mid-ranks if ties occur – and perform the intended analysis on these. It will often be the case that the assumptions will be closer to what is required. On the other hand, the hypotheses will now be about the ranks rather than the original data. Also the reader should be aware there are caveats concerning the rank transform method. See, for example, https://en.wikipedia.org/wiki/ANOVA_on_ranks.

The second example is in archetypical CMH format. We return to the strawberry data in Chapters 8 and 10 and to the homosexual marriage data in Chapters 2, 3, 6, 7, and 11.

1.4 Examples

1.4.1 Strawberry Data

The data in Table 1.1 are from Pearce (1960). Pesticides are applied to strawberry plants to inhibit the growth of weeds. The response represents the total spread in inches of twelve plants per plot approximately two months after the application of the weedkillers. The question is, do they also inhibit the growth of the strawberries? Pesticide O is a control. The design is a supplemented balanced design.

The means for pesticides A–O are 142.4, 144.4, 103.6, 123.8, and 174.625 respectively. Pearce (1960) found a strong treatment effect: the pesticides do appear to inhibit the growth of the strawberries.

Using ®, a pesticides p-value of less than 0.0001 is found. Blocks have a p-value of 0.0304 indicating blocking is important. The Shapiro–Wilk test applied to the residuals gave a p-value of 0.1313. There appears to be no problem with the parametric model.

Table 1.1   Growth of strawberry plants after applying pesticides.

Block I	Block II	Block III	Block IV
C, 107 (5)	A, 136 (14)	B, 118 (8)	O, 173 (23)
A, 166 (21.5)	O, 146 (16)	A, 117 (7)	C, 95 (1)
D, 133 (13)	C, 104 (4)	O, 176 (24)	C, 109 (6)
B, 166 (21.5)	B, 152 (18)	D, 132 (11.5)	A, 130 (10)
O, 177 (25)	D, 119 (9)	B, 139 (15)	D, 103 (2.5)
A, 163 (19)	O, 164 (20)	O, 186 (27)	O, 185 (26)
O, 190 (28)	D, 132 (11.5)	C, 103 (2.5)	B, 147 (17)

```
lm_strawberry = lm(response~pesticide+block)
Anova(lm_strawberry, type=3)

## Anova Table (Type III tests)
##
## Response: response
##              Sum Sq Df  F value     Pr(>F)
## (Intercept)  87318  1 504.6923 1.161e-15 ***
## pesticide    17029  4  24.6072 1.757e-07 ***
## block         1824  3   3.5142   0.03405 *
## Residuals     3460 20
## ---
## Signif. codes:  0 '***' 0.001 '**' 0.01 '*' 0.05 '.' 0.1 ' ' 1

shapiro.test(lm_strawberry$residuals)
```

```
##
##      Shapiro-Wilk normality test
##
## data:  lm_strawberry$residuals
## W = 0.94294, p-value = 0.1313
```

Nevertheless the same analysis was applied to the ranked data that are given in parentheses in Table 1.1. A pesticides *p*-value of less than 0.0001 is found. Blocks have a *p*-value of 0.0308 indicating blocking is important. The Shapiro–Wilk test applied to the residuals gave a *p*-value of 0.0909. There appears to be no problem with this model.

The two analyses are almost identical.

```
lm_strawberry = lm(rank~pesticide+block)
anova(lm_strawberry)

## Analysis of Variance Table
##
## Response: rank
##             Df  Sum Sq Mean Sq F value     Pr(>F)
## pesticide    4 1384.33  346.08  24.225 1.996e-07 ***
## block        3  155.45   51.82   3.627   0.03075 *
## Residuals   20  285.73   14.29
## ---
## Signif. codes:  0 '***' 0.001 '**' 0.01 '*' 0.05 '.' 0.1 ' ' 1

shapiro.test(lm_strawberry$residuals)

##
##      Shapiro-Wilk normality test
##
## data:  lm_strawberry$residuals
## W = 0.93665, p-value = 0.09086
```

1.4.2   Homosexual Marriage Data

Scores of 1, 2, and 3 are assigned to the responses agree, neutral, and disagree respectively to the proposition 'Homosexuals should be able to marry' and scores of 1, 2, and 3 are assigned to the religious categories fundamentalist, moderate, and liberal respectively. See Table 1.2. Pearson tests on the overall table and on the table obtained by aggregating the strata School and College both have *p*-values 0.000. There is strong evidence of an association between the proposition responses and religion, but perhaps there is more information in the data. Analysis from Agresti (2003) finds the three CMH tests have *p*-values 0.000, agreeing with the conclusion that there is strong evidence of an association between the proposition responses and religion. The ordinal

Table 1.2 Opinions on homosexual marriage by religious beliefs and education levels for ages 18–25.

Education	Religion	Homosexuals should be able to marry		
		Agree	Neutral	Disagree
School	Fundamentalist	6	2	10
	Moderate	8	3	9
	Liberal	11	5	6
College	Fundamentalist	4	2	11
	Moderate	21	3	5
	Liberal	22	4	1

CMH tests that we will describe in Chapter 2 find evidence of mean differences in the responses and of a (linear–linear) correlation between responses and religion, corroborating the CMH inference.

The parametric analysis finds the mean responses for fundamentalist, moderate, and liberal are 2.314, 1.694, and 1.469, respectively. Using ®, a religion p-value of 0.000 is found. Education (blocks) has a p-value of 0.011 indicating blocking is important. The Shapiro–Wilk test applied to the residuals gave a p-value of 0.000. The parametric model is problematic.

```
lm_marriage_para = lm(as.numeric(opinion)~religion+education)
Anova(lm_marriage_para, type = 3)

## Anova Table (Type III tests)
##
## Response: as.numeric(opinion)
##                 Sum Sq  Df  F value    Pr(>F)
## (Intercept) 173.895    1 255.9338 < 2.2e-16 ***
## religion     14.022    2  10.3186 6.966e-05 ***
## education     4.506    1   6.6316   0.01115 *
## Residuals    87.649 129
## ---
## Signif. codes:  0 '***' 0.001 '**' 0.01 '*' 0.05 '.' 0.1 ' ' 1


shapiro.test(lm_marriage_para$residuals)

##
##      Shapiro-Wilk normality test
##
## data:  lm_marriage_para$residuals
## W = 0.94108, p-value = 2.02e-05
```

Applying the same analysis to the mid-ranks of the response categories finds the mean responses for fundamentalist, moderate, and liberal are 87.30, 63.79, and 55.71, respectively. Using ®, a religion *p*-value of 0.000 is found. Education (blocks) has a *p*-value of 0.010 indicating blocking is important. The Shapiro–Wilk test applied to the residuals gave a *p*-value of 0.000. The parametric model is again problematic.

In both examples the two analyses are almost identical.

```
lm_marriage_rank = lm(rank(opinion)~religion+education)
Anova(lm_marriage_rank, type = 3)

## Anova Table (Type III tests)
##
## Response: rank(opinion)
##               Sum Sq  Df  F value     Pr(>F)
## (Intercept)   248562    1 246.2417 < 2.2e-16 ***
## religion       19618    2   9.7177 0.0001172 ***
## education       6817    1   6.7538 0.0104435 *
## Residuals     130216  129
## ---
## Signif. codes:  0 '***' 0.001 '**' 0.01 '*' 0.05 '.' 0.1 ' ' 1

shapiro.test(lm_marriage_rank$residuals)

##
##      Shapiro-Wilk normality test
##
## data:  lm_marriage_rank$residuals
## W = 0.93588, p-value = 8.703e-06
```

Chapter 2 describes the basic CMH tests in detail.

Bibliography

Agresti, A. (2003). *Categorical Data Analysis*. Hoboken, NJ: John Wiley & Sons.

Conover, W. J. and Iman, R. L. (1981). Rank transformations as a bridge between parametric and nonparametric statistics. *The American Statistician*, 35(3):124–129.

Pearce, S. C. (1960). Supplemented balance. *Biometrika*, 47(3/4):263–271.

Rayner, J. C. W. and Best, D. J. (1999). Modelling ties in the sign test. *Biometrics*, 55(2):663–665.

2

The Basic CMH Tests

2.1 Genesis: Cochran (1954), and Mantel and Haenszel (1959)

The origins of the Cochran–Mantel–Haenszel (CMH) methodology go back to Cochran (1954), a wide-ranging paper about the use of the Pearson X^2 tests of goodness of fit and of association, and Mantel and Haenszel (1959), which focused on methodological issues. In the former a test for average partial association was proposed, while the latter developed the test using a hypergeometric model. A quarter-century of development was consolidated by Landis and his colleagues in two important papers: the review of Landis et al. (1978) and the description of the computer program PARCAT in Landis et al. (1979). PARCAT implemented the CMH methodology and the program was made available to users 'for a modest cost'. Landis et al. (1978) is our starting point, the basis of the methodology described in subsequent sections of this chapter. But first we review the genesis of that methodology.

First, we need to discuss notation. We use X^2 for the test statistic and χ^2 for the distribution. We use subscripts on X^2 to distinguish between variants of the test statistics, such as X_P^2 for the Pearson statistic that involves no parameter estimation and X_{PF}^2 for the Pearson–Fisher statistic that involves maximum likelihood estimation of the unknown parameters when the data are categorical. There are several other named X^2 tests. Some of the early papers, Cochran (1954) included, used χ^2 where we would use X^2.

The title of Cochran (1954) announces that the paper is about 'strengthening the common χ^2 tests': the Pearson tests of goodness of fit and of association. It begins by observing that these are omnibus tests, and hence cannot be expected to be as powerful in detecting alternatives of specified

An Introduction to Cochran–Mantel–Haenszel Testing and Nonparametric ANOVA, First Edition. J.C.W. Rayner and G. C. Livingston Jr.
© 2023 John Wiley & Sons Ltd. Published 2023 by John Wiley & Sons Ltd.

interest as more focused and alternative tests. Subsequently he discusses several tests based on components.

There is a section that addresses strengthening the Pearson X^2 tests, where he revisits advice on the issue of small cell expectations. Early studies had indicated that the approximation of the χ^2 distribution to the null distribution of X^2 was poor unless the cell expectations all exceeded 5. Cochran felt that was conservative, and resulted in excessive pooling, especially in the tails. An example is given when such pooling causes a substantial loss of power. Recommendations for minimum cell expectations are given.

Sections 3, 4, and 5 concern goodness of fit for the Poisson, binomial and normal distributions and consider using components of X^2 and alternative tests to test for specific alternatives. Section 6 is titled *Subdivision of degrees of freedom in the 2 × N contingency table*. In a similar theme the test statistic X^2 is expressed as a weighted sum of squares, which is subdivided in various ways to obtain sets of independent components to test various aspects of the null hypothesis. This is followed by consideration of the general two-way contingency table. The approach uses techniques pioneered by Irwin (1949) and Lancaster (1949) to partition X^2 into additive components, each with a single degree of freedom.

Of greatest interest to the formulation of the CMH methodology is the next section, Section 8, *The combination of 2 × 2 contingency tables*. Cochran first considers aggregating multiple 2 × 2 tables into a single 2 × 2 table. Next he considers calculating X^2 for each table, and, since tables are assumed to be independent, adding the X^2 values and using the asymptotic χ^2 distribution with degrees of freedom the number of tables. Potential weaknesses in this approach are discussed and examples given.

The aggregating tables approach is deemed poor, since differences in proportions in individual tables may sometimes be positive, and sometimes negative, and hence cancel when aggregated. To overcome this issue Cochran suggests aggregating X values, taking into account the signs of the proportion differences. The test statistic, suitably standardised, is referred to the standard normal tables.

Finally, in the words of Landis et al. (1978)

> Cochran (1954) proposed a test directed at "average partial associa-
> tion"(Landis' words, not Cochran's) for a set of (2×2) tables using a
> mean difference weighted across the q tables determined by the levels
> of the covariables. For this case, he recommended a test statistic based
> on asymptotic binomial model results which require moderately large
> sample sizes (e.g. $nh > 20$) for each of the tables.

A plausibility argument is given.

The effect of extraneous variation is briefly addressed in Section 9, that concludes with the warning 'A thorough discussion of when not to use X^2 tests in problems of this kind, and of the best alternatives, would be lengthy.' Critical thinking about the suitability of an X^2 test is encouraged.

In the paper's conclusion Cochran advises that 'the most useful principle is to think in advance about the way in which the data seem likely to depart from the null hypothesis. This often leads to the selection of a single test ... as the only one that appears appropriate.' A warning is given to not apply multiple tests of significance to the one data set.

It is interesting that from such a wide-ranging paper one issue generated substantial interest. One consequence of Mantel and Haenszel (1959) is that the analysis of a set of 2×2 tables can be approached using a hypergeometric model which either permits exact tests or requires only the overall sample size to be reasonably large for asymptotic methods to be applicable.

Mantel and Haenszel (1959) considers the design and analysis of data from retrospective studies of disease. To understand why the hypergeometric model is appropriate, it is necessary to understand what a retrospective study is. In *retrospective studies*, individuals are sampled and information is collected about their past. ...In *prospective studies*, individuals are followed over time and data about them is collected as their characteristics or circumstances change. Mantel and Haenszel (1959), say:

> A retrospective study of disease occurrence may be defined as one in which the determination of association of a disease with some factor is based on an unusually high or low frequency of that factor among diseased persons. This contrasts with a forward study in which one looks instead for an unusually high or low occurrence of the disease among individuals possessing the factor in question.

They are interested in data of the form shown in Table 2.1.

Table 2.1 The style of contingency table addressed.

	With factor	Without factor	Total
Diseased	A	B	N_1
Disease free (controls)	C	D	N_2
Total	M_1	M_2	T

Source: Adapted from Mantel and Haenszel (1959).

They note that a commonly employed statistical test of association is the X^2 test on the difference between the cases and controls in the proportion of individuals having the factor under test. A corrected X^2 may be calculated routinely as

$$\frac{\left(|AD - BC| - \frac{T}{2}\right)^2 T}{M_1 M_2 N_1 N_2}$$

and tested by referring X^2 to χ_1^2 in the usual manner.

In the section *Statistical Procedures for Factor Control* the use of subcategories (strata) is discussed. In each stratum tables like the previous are calculated. In calculating a corrected chi-squared test statistic on each stratum the authors assume the marginal totals are known and hence use moments of the hypergeometric distribution, so the test statistic is

$$\frac{\left(|AD - BC| - \frac{T}{2}\right)^2 (T - 1)}{M_1 M_2 N_1 N_2}.$$

Although the difference is only the factor $\frac{T-1}{T}$ in each stratum, if the stratum totals are not large, this may make a considerable difference in the test of significance.

In a prospective (forward) study, within each stratum the numbers with and without the factor (M_{1j} and M_{2j}) may well be known, it is most unlikely that this will be true for the numbers with and without the disease (N_{1j} and N_{2j}). However, it is feasible that in a retrospective study the latter will be known.

In 1978 the methodology hinted at here had been extensively developed and generalised. The description in Landis et al. (1978), given in the next section, summarises these results.

2.2 The Basic CMH Tests

The CMH tests are a class of tests used to nonparametrically analyse tables N_{ihj} of count data of a particular structure. Specifically, N_{ihj} counts the number of observations that are classified as being treatment i, falling into outcome category h, and in the jth stratum, $i = 1, \ldots, t$, $h = 1, \ldots, c$, and $j = 1, \ldots, b$. There are at least two treatments, at least two outcome categories and strata are independent. We refer to this design subsequently as the *CMH design*. The null hypothesis of interest is that having adjusted for the b strata, there is no association between the treatment and response variables. Using four basic tests the null hypothesis is tested against various alternatives that will be explored subsequently. As is usual, note that N_{ihj} is a random variable and n_{ihj} is a particular value of that random variable.

Subsequently the dot notation is used to reflect summation over a subscript. In the traditional CMH tests (see, for example, Landis et al. (1978), Landis et al. (1979), Davis (2002), and Kuritz et al. (1988))

- the strata totals $\sum_{i=1}^{t} \sum_{h=1}^{c} N_{ihj} = n_{\bullet\bullet j}$,
- the treatment totals within strata, $\sum_{h=1}^{c} N_{ihj} = n_{i\bullet j}$, and
- the outcome totals within strata, $\sum_{i=1}^{t} N_{ihj} = n_{\bullet hj}$

are not random variables. Effectively they are all known prior to sighting the data. Inference is *conditional* on the marginals taking these values.

The traditional CMH tests assess

- overall partial association (OPA): S_{OPA} is asymptotically distributed as $\chi^2_{b(c-1)(t-1)}$,
- general association (GA): S_{GA} is asymptotically distributed as $\chi^2_{(c-1)(t-1)}$,
- mean scores (MS): S_{MS} is asymptotically distributed as $\chi^2_{(t-1)}$, and
- correlation (C): S_{C} is asymptotically distributed as χ^2_1.

The aforementioned provides the symbols used subsequently for each test statistic and the asymptotic null distributions of the test statistics. The test statistics are quadratic forms with vectors using the elements in the table counts. Some of the covariance matrices involve Kronecker products; see the Appendix A.1 for mathematical background on Kronecker products.

The degrees of freedom of the approximating chi-squared distributions of the CMH OPA, GA, MS, and C statistics are $b(c-1)(t-1)$, $(c-1)(t-1)$, $(t-1)$, and 1, respectively: strictly decreasing. This reflects the fact that the alternatives to the null hypothesis are becoming increasingly stringent. The CMH OPA test assesses *overall* association while the CMH GA test seeks to detect an *average* association between the treatment and response variables. The CMH MS test seeks to detect treatment *mean* differences between the responses and the CMH C test seeks to detect *linear association* between the treatment and response variables.

An analogy might be helpful here. Suppose a doctor has just 30 minutes to check 30 people for a disease. If every person is assessed, a one minute scrutiny may fail to diagnose an ill patient. On the other hand, checking just six people will permit a more thorough assessment of those six, but cannot give useful information about the 24 people not assessed. Finally, a complete assessment of just one patient may determine with certainty if that patient has the disease, but says nothing about the remainder.

The tests with larger degrees of freedom are more *omnibus*, detecting general alternatives with moderate power. As the degrees of freedom decrease, the tests become more *focused*, having greater power in the reduced parameter space but being insensitive to alternatives outside of that space. Thus, for detecting correlation effects the correlation test will have greater power than GA test, but it will have less power than the GA test for other alternatives.

The CMH design is appropriate for, for example, categorical randomised block data. While the CMH methodology is appropriate more generally, it won't accommodate more complex designs such as Latin square, multifactor ANOVA and many other designs. However, it is an extremely important analysis tool for randomised blocks when the responses are categorical rather than continuous. In consumer studies with *just about right* (JAR) responses such as in the Jams Example following, when the number of response categories is small the randomised block F test may be invalid in spite of the well-known ANOVA robustness.

To give a flavour of the sort of data sets to which the CMH methodology applies, two examples will now be considered: details of the tests will be given subsequently.

2.2.1 Homosexual Marriage Data

This data set was looked at in detail in Section 1.4 using the data presented in Table 1.2. Agresti (2002) finds S_{GA} takes the value 19.76 with χ_4^2 p-value equal to 0.0006, S_{MS} takes the value 17.94 with χ_2^2 p-value equal to 0.0001, and S_C takes the value 16.83 with χ_1^2 p-value less than 0.0001. The sampling distributions of the test statistics are all asymptotically χ^2. Here $n_{\bullet\bullet\bullet} = 133$, so the asymptotic distributions should be acceptable.

The more omnibus OPA test finds evidence of associations between the variables. In some data sets the parameter space, as indicated by the degrees of freedom, is so large that the test has low power, and may miss evidence of associations. It is then reasonable to aggregate strata and apply the GA test. This sacrifices finding associations between strata with the goal of better assessing if there are average associations between responses and religion. That isn't necessary here, but it would be useful to know something about what those associations are. The MS test finds evidence of differences in the mean responses, while the C test finds evidence of a (linear–linear) correlation between responses and religion.

We will analyse this data set more deeply subsequently, but it should be clear that applying the suite of CMH tests gives a deeper assessment of a data set than just using Pearson tests.

2.2.2 Jams Data

Three plum jams, A, B, and C are given *JAR* sweetness codes by eight judges as in the Table 2.2. Here 1 denotes *not sweet enough*, 2 *not quite sweet enough*,

Table 2.2 Three jams ranked for sweetness by eight judges.[a]

Jam	Blocks (judges)							
	1	2	3	4	5	6	7	8
1	3 (2.5)	4 (1.5)	3 (2.5)	1 (1)	2 (1.5)	1 (1)	2 (1)	2 (1.5)
2	2 (1)	5 (3)	2 (1)	4 (3)	4 (3)	3 (2.5)	5 (3)	5 (3)
3	3 (2.5)	4 (1.5)	3 (2.5)	2 (2)	2 (1.5)	3 (2.5)	4 (2)	2 (1.5)

a) Low JAR scores were ranked 1. The ranked data are in parentheses.

3 *just about right*, 4 *a little too sweet*, and 5 *too sweet*. The data was also analysed in Rayner and Best (2017). The ranks are included and will be used when the data is revisited in Section 4.3. To calculate the ordinal CMH tests we need *scores*: values assigned to classes as being representative of the observations in those classes. The scores 1, 2, and 3 are assigned to jams A, B, and C, respectively, perhaps based on increasing sugar content, or even cost. The sweetness codes are used as response scores. There are other options, such as mid-ranks, but these choices are natural for this data set. The treatment sums for jams A, B, and C are 18, 30, and 23, respectively. We find $S_{MS} = 9.6177$ with χ_2^2 p-value 0.0082 and $S_C = 1.1029$ with χ_1^2 p-value 0.2936. There is evidence of a mean effect: on average the jams are different. However, there is no evidence of a correlation effect: that as we pass from jam A to B and then to C there is no evidence of a linearly increasing (or decreasing) response.

2.3 The Nominal CMH Tests

The CMH MS test assumes that response categories are ordered while the CMH C test assumes that both treatment categories and response categories are ordered. We have labelled these as *ordinal* CMH tests, while the OPA and GA CMH tests, that make no assumption about ordering, can be called *nominal* CMH tests. Scores for the ordered categories are needed to apply the ordinal tests, while no scores are required for the nominal tests.

It is important to note that the traditional CMH tests are conditional tests, conditional on the treatment totals within strata, $\sum_{h=1}^{c} N_{ihj} = n_{i \bullet j}$, and the outcome totals within strata, $\sum_{i=1}^{t} N_{ihj} = n_{\bullet hj}$, being known prior to collecting or sighting the data. For randomised block data in which the responses are untied ranks the marginal totals are known before collecting the data. This is because each treatment is applied once in every block and each response is a rank that is assigned once. This is not the case for data such as in the Homosexual Marriage data.

The CMH test statistics are all quadratic forms of the form $(X - \mathrm{E}[X])^{\mathrm{T}}$ $\mathrm{Cov}(X)^{-1}(X - \mathrm{E}[X])$ in which X involves the N_{ihj} stacked into a vector. To calculate $\mathrm{E}[X]$ and $\mathrm{Cov}(X)$ distributional assumptions must be made. For each stratum since the row and column totals are known, the counts $\{N_{ihj}\}$ can be assumed to follow an extended hypergeometric distribution. Moreover since the strata are mutually independent, a product distribution is appropriate for the collection of these strata counts. The probability function for $\{N_{ihj}\}$ is

$$P\left(N_{ihj} = n_{ihj} \text{ for all } i, h, \text{ and } j\right) = \frac{\prod_{i=1}^{t} n_{i\bullet j}! \; \prod_{h=1}^{c} n_{\bullet hj}!}{n! \prod_{i=1}^{t} \prod_{h=1}^{c} n_{ihj}!}.$$

To define the CMH GA test statistic S_{GA}, first define the vector of counts on the jth stratum $U_j = (N_{11j}, \ldots, N_{1cj}, \ldots, N_{t1j}, \ldots, N_{tcj})^{\mathrm{T}}$. Summing over strata gives

$$U_\bullet = (N_{11\bullet}, \ldots, N_{1c\bullet}, \ldots, N_{t1\bullet}, \ldots, N_{tc\bullet})^{\mathrm{T}}.$$

Using the product extended hypergeometric distribution, it may be shown, as in Landis et al. (1978), that $U_\bullet$ has mean with typical element

$$\mathrm{E}\left[N_{ih\bullet}\right] = \sum_{j=1}^{b} \frac{n_{i\bullet j} n_{\bullet hj}}{n_{\bullet\bullet j}}$$

and covariance matrix

$$\mathrm{Cov}\left(U_\bullet\right) = \sum_{j=1}^{b} \frac{n_{\bullet\bullet j}^2}{n_{\bullet\bullet j} - 1} \left(V_{\mathrm{T}j} \otimes V_{\mathrm{C}j}\right)$$

in which, after writing $p_{i\bullet j} = \frac{n_{i\bullet j}}{n_{\bullet\bullet j}}$ and $p_{\bullet hj} = \frac{n_{\bullet hj}}{n_{\bullet\bullet j}}$,

$$V_{\mathrm{T}j} = \left\{ \mathrm{diag}\left(p_{i\bullet j}\right) - \left(p_{i\bullet j}\right)\left(p_{i\bullet j}\right)^{\mathrm{T}} \right\}, \text{ and}$$

$$V_{\mathrm{C}j} = \left\{ \mathrm{diag}\left(p_{\bullet hj}\right) - \left(p_{\bullet hj}\right)\left(p_{\bullet hj}\right)^{\mathrm{T}} \right\}.$$

Here $\otimes$ is the Kronecker or direct product.

The CMH GA test statistic is a quadratic form using $U_\bullet - \mathrm{E}[U_\bullet]$ and an inverse of $\mathrm{Cov}(U_\bullet)$, namely, $S_{\mathrm{GA}} = (U_\bullet - \mathrm{E}[U_\bullet])^{\mathrm{T}} \mathrm{Cov}^-(U_\bullet)$ $(U_\bullet - \mathrm{E}[U_\bullet])$. As $\mathrm{Cov}(U_\bullet)$ is not of full rank, either a generalised inverse $\mathrm{Cov}^-(U_\bullet)$ can be used, or dependent variables can be chosen and omitted to produce a covariance matrix of full rank. See Appendix A.2 for background information on the Moore–Penrose generalised inverse. All unknown parameters are estimated using maximum likelihood under the null hypothesis. Asymptotically, as $n_{\bullet\bullet\bullet}$ becomes large, S_{GA} has the $\chi^2_{(c-1)(t-1)}$ distribution. The test statistic is symmetric in the treatments and outcome categories, and independent of the choice of the dependent variables.

The test statistic S_{GA} is usually too complicated for routine hand calculation; it is almost always applied using software in packages such as ℝ. However, in Chapters 3 and 4 we will give simple forms for the completely randomised and randomised block designs (RBDs).

To calculate S_{OPA} the vector of the quadratic form involves the aggregation of the U_j via $U = \left(U_1^T, \ldots, U_b^T \right)^T$. The covariance matrix is again calculated using the product extended hypergeometric distribution. To give a simple expression for S_{OPA}, first define the Pearson statistic on the jth block:

$$X_{Pj}^2 = \sum_{i=1}^{t} \sum_{h=1}^{c} \frac{\left(N_{ihj} - \frac{n_{i\bullet j} n_{\bullet hj}}{n_{\bullet\bullet j}} \right)^2}{\frac{n_{i\bullet j} n_{\bullet hj}}{n_{\bullet\bullet j}}}.$$

The CMH OPA test statistic is

$$S_{OPA} = \sum_{j=1}^{b} \left(\frac{n_{\bullet\bullet j} - 1}{n_{\bullet\bullet j}} \right) X_{Pj}^2 = \sum_{j=1}^{b} \left(\frac{n_{\bullet\bullet j} - 1}{n_{\bullet\bullet j}} \right) \sum_{i=1}^{t} \sum_{h=1}^{c} \frac{\left(N_{ihj} - \frac{n_{i\bullet j} n_{\bullet hj}}{n_{\bullet\bullet j}} \right)^2}{\frac{n_{i\bullet j} n_{\bullet hj}}{n_{\bullet\bullet j}}},$$

the sum over strata of weighted strata Pearson X^2 statistics, the weight on the jth stratum being $\frac{n_{\bullet\bullet j} - 1}{n_{\bullet\bullet j}}$. Asymptotically this has the $\chi^2_{b(c-1)(t-1)}$ distribution.

One difference between the two tests is that the CMH GA test is seeking to detect *average* partial association while the CMH OPA test is seeking to detect *overall* partial association. The former is more *focused*, with $(c-1)(t-1)$ degrees of freedom, compared with $b(c-1)(t-1)$ for the more *omnibus* CMH OPA statistic. The degrees of freedom are the dimension of the parameter space specified by the alternative hypotheses. Thus the CMH OPA test is seeking to detect very general alternatives to the null hypothesis and will have relatively low power for these alternatives. The CMH GA test seeks to detect fewer alternatives and will have more power than the CMH OPA test for these alternatives. However, CMH OPA test will have some power for alternatives to which the CMH GA is insensitive. In other contexts more focused tests have been constructed using components of an omnibus test statistic.

An alternative test for OPA is the Pearson test, with statistic

$$X_P^2 = \sum_{i=1}^{t} \sum_{h=1}^{c} \sum_{j=1}^{b} \frac{\left(N_{ihj} - \frac{n_{i\bullet j} n_{\bullet hj}}{n_{\bullet\bullet j}} \right)^2}{\frac{n_{i\bullet j} n_{\bullet hj}}{n_{\bullet\bullet j}}}.$$

This is an unconditional test that does not assume all treatment and outcome categories are known before sighting the data. The difference between

S_{OPA} and X_{P}^2 is merely the presence or absence of the weights $\frac{n_{\bullet\bullet j}-1}{n_{\bullet\bullet j}}$. For large stratum counts this will make little difference in the values of S_{OPA} and X_{P}^2.

Likewise GA can be tested for using a Pearson test for the two-way table of counts $\{N_{ih\bullet}\}, X_{\text{P}\bullet}^2$, say. The Pearson test statistics X_{P}^2 and $X_{\text{P}\bullet}^2$ have the same asymptotic distributions as the corresponding conditional tests. Most users will have more familiarity with the unconditional tests and most packages will have routines for their calculation even if they don't have routines for CMH tests.

Subsequently our main focus will be on the CMH ordinal tests.

2.4 The CMH Mean Scores Test

Suppose, as before, that N_{ihj} counts the number of times an observation is classified as being treatment i falling into outcome category h in the jth stratum, $i = 1, \ldots, t, h = 1, \ldots, c$, and $j = 1, \ldots, b$. Assume that outcomes are ordinal and assign the *score* b_{hj} to the hth response on the jth stratum. All marginal totals are assumed to be known, so the product extended hypergeometric model is assumed. The score sum for treatment i in stratum j is $M_{ij} = \sum_{h=1}^c b_{hj} N_{ihj}$. Taking expectations $\text{E}\left[M_{ij}\right] = n_{i\bullet j} \sum_{h=1}^c b_{hj} \frac{n_{\bullet hj}}{n_{\bullet\bullet j}}$ since $\text{E}\left[N_{ihj}\right] = \frac{n_{i\bullet j} n_{\bullet hj}}{n_{\bullet\bullet j}}$. Inference is based on $M = \{M_{i\bullet}\}$ through the quadratic form

$$S_{\text{MS}} = (M - \text{E}[M])^{\text{T}} \text{Cov}^- (M) (M - \text{E}[M]),$$

in which all unknown parameters are estimated by maximum likelihood under the null hypothesis. As strata are independent, $\text{Cov}(M) = \sum_{j=1}^b \text{Cov}(M_j)$. To give $\text{Cov}(M_j)$ first define

$$S_j^2 = \frac{n_{\bullet\bullet j} \sum_{h=1}^c b_{hj}^2 n_{\bullet hj} - \left(\sum_{h=1}^c b_{hj} n_{\bullet hj}\right)^2}{n_{\bullet\bullet j} - 1}.$$

Then, as shown next, $\text{Cov}(M_j) = S_j^2 V_{\text{T}j}$. Since $\text{Cov}(M_j)$ is not of full rank the usual approaches, such as dropping appropriate treatment and/or outcome categories, or using a generalised inverse, can be used.

Note: The MS statistic depends on the scores assigned to the response categories. The test statistic could therefore be written $S_{\text{MS}}(\{b_{hj}\})$ to emphasise this dependence.

The derivation of $\text{Cov}(M_j)$ requires routine but tedious algebra. If δ_{uv} is the Kronecker delta, which takes the value 1 if $u = v$ and zero otherwise, using standard distribution theory for the product extended hypergeometric distribution, $E\left[N_{ihj}\right] = \frac{n_{i\bullet j}n_{\bullet hj}}{n_{\bullet\bullet j}}$ and the covariance between N_{ihj} and $N_{i'h'j}$ is

$$\frac{n_{i\bullet j}n_{\bullet hj}\left(\delta_{ii'}n_{\bullet\bullet j} - n_{i'\bullet j}\right)\left(\delta_{hh'}n_{\bullet\bullet j} - n_{\bullet h'j}\right)}{\left(n_{\bullet\bullet j}^2\left(n_{\bullet\bullet j} - 1\right)\right)}.$$

It follows that

$$\text{Var}\left(\sum_{h=1}^{c}b_{hj}N_{ihj}\right) = \frac{n_{i\bullet j}\left(n_{\bullet\bullet j} - n_{i\bullet j}\right)S_j^2}{\left(n_{\bullet\bullet j}^2\left(n_{\bullet\bullet j} - 1\right)\right)}$$

and

$$\text{Cov}\left(\sum_{h=1}^{c}b_{hj}N_{ihj}, \sum_{h'=1}^{c}b_{h'j}N_{i'h'j}\right) = \frac{-n_{i\bullet j}n_{i'\bullet j}S_j^2}{\left(n_{\bullet\bullet j}^2\left(n_{\bullet\bullet j} - 1\right)\right)}.$$

Aggregating these results, if $V = \left\{M_{i\bullet} - \sum_{j=1}^{b}n_{i\bullet j}\sum_{h=1}^{c}\frac{b_{hj}n_{\bullet hj}}{n_{\bullet\bullet j}}\right\}$ it follows that S_{MS} can be calculated from

$$V^{\text{T}}\left[\sum_{j=1}^{b}S_j^2\left(\text{diag}\left(p_{i\bullet j}\right) - \left\{p_{i'\bullet j}p_{i\bullet j}\right\}\right)\right]^{-}V.$$

Now put $D_j = \text{diag}\left(\sqrt{p_{i\bullet j}}\right)$ and $u_j = \left\{\sqrt{p_{i\bullet j}}\right\}$. Then $\text{Cov}\left(M_j\right) = S_j^2 D_j\left(I_t - u_j u_j^{\text{T}}\right)D_j$. In some circumstances this form of $\text{Cov}\left(M_j\right)$ will be more convenient.

Under the null hypothesis of no treatment effects the distribution of S_{MS} can be shown to be asymptotically distributed as χ_{t-1}^2; see, for example, Landis et al. (1978). One way to show this is to note the statistic is of the form $X^{\text{T}}V^{-}X$ in which, by the central limit theorem (CLT), X is asymptotically multivariate normal, $V = \text{Cov}(X)$, so that $X^{\text{T}}V^{-}X$ is asymptotically χ^2. The degrees of freedom are the rank of V, which is

$$\text{rank}\,(V) = \text{rank}\left(S_j^2 D_j\left(I_t - u_j u_j^{\text{T}}\right)D_j\right) = \text{rank}\left(I_t - u_j u_j^{\text{T}}\right)$$

$$= \text{trace}\left(I_t - u_j u_j^{\text{T}}\right) = t - \text{trace}\left(u_j u_j^{\text{T}}\right) = t - \text{trace}\left(u_j^{\text{T}} u_j\right)$$

$$= t - 1.$$

2.5 The CMH Correlation Test

2.5.1 The CMH C Test Defined

The CMH correlation tests assume that the treatment and response variables are both measured on either an ordinal or the interval scale, and that for the ith treatment the scores for the hth category are a_{hi}, $i = 1, \ldots, t$, and on the jth stratum the response scores for the hth category are $b_{hj}, j = 1, \ldots, b$, both for $h = 1, \ldots, c$.

The null hypothesis is that there is no association between the treatment and response variables after adjusting for the b strata. This is tested against the alternative that across strata there is a consistent linear association, positive or negative, between the treatment scores and response scores.

Take

- $C_j = \sum_{i=1}^{t} \sum_{h=1}^{c} a_{ij} b_{hj} \left(N_{ihj} - \mathrm{E}\left[N_{ihj}\right]\right)$, and
- $C = \sum_{j=1}^{b} C_j$.

The CMH correlation (C) statistic is $\frac{C^2}{\mathrm{Var}(C)} = S_{\mathrm{C}}$, say. The derivation of $\mathrm{Var}\,(C)$ is relatively complex if scalars are used, but is routine using Kronecker products.

To derive $\mathrm{Var}\,(C)$ first define $a_j = \left(a_{1j}, \ldots, a_{tj}\right)^{\mathrm{T}}$, $b_j = \left(b_{1j}, \ldots, b_{cj}\right)^{\mathrm{T}}$, and $N_j = \left(N_{11j}, \ldots, N_{1cj}, \ldots, N_{t1j}, \ldots, N_{tcj}\right)^{\mathrm{T}}$. Then $C_j = \left(a_j \otimes b_j\right)^{\mathrm{T}} \left(N_j - \mathrm{E}\left[N_j\right]\right)$ and

$$\mathrm{Var}\,(C_j) = \mathrm{E}\left[\left(a_j \otimes b_j\right)^{\mathrm{T}} \left(N_j - \mathrm{E}\left[N_j\right]\right)\left(N_j - \mathrm{E}\left[N_j\right]\right)^{\mathrm{T}} \left(a_j \otimes b_j\right)\right]$$

$$= \left(a_j \otimes b_j\right)^{\mathrm{T}} \mathrm{E}\left[\left(N_j - \mathrm{E}\left[N_j\right]\right)\left(N_j - \mathrm{E}\left[N_j\right]\right)^{\mathrm{T}}\right]\left(a_j \otimes b_j\right)$$

$$= \left(a_j \otimes b_j\right)^{\mathrm{T}} \mathrm{Cov}\,\left(N_j\right)\left(a_j \otimes b_j\right).$$

Recall from Section 2.3 that with $p_{i\bullet j} = \frac{n_{i\bullet j}}{n_{\bullet\bullet\bullet}}$ and $p_{\bullet hj} = \frac{n_{\bullet hj}}{n_{\bullet\bullet\bullet}}$, we have $V_{\mathrm{T}j} = \mathrm{diag}\,\left(p_{i\bullet j}\right) - \left(p_{i\bullet j}\right)\left(p_{i\bullet j}\right)^{\mathrm{T}}$ and $V_{\mathrm{C}j} = \mathrm{diag}\,\left(p_{\bullet hj}\right) - \left(p_{\bullet hj}\right)\left(p_{\bullet hj}\right)^{\mathrm{T}}$. Now from Landis et al. (1978)

$$\mathrm{Cov}\,\left(N_j\right) = \frac{n_{\bullet\bullet j}^2}{n_{\bullet\bullet j} - 1}\left(V_{\mathrm{T}j} \otimes V_{\mathrm{C}j}\right).$$

Hence,

$$\mathrm{Var}\,(C_j) = \frac{n_{\bullet\bullet j}^2}{n_{\bullet\bullet j}}\left(a_j^{\mathrm{T}} V_{\mathrm{T}j} a_j \otimes b_j^{\mathrm{T}} V_{\mathrm{C}j} b_j^{\mathrm{T}}\right) = \frac{n_{\bullet\bullet j}^2}{n_{\bullet\bullet j}}\left(a_j^{\mathrm{T}} V_{\mathrm{T}j} a_j\right)\left(b_j^{\mathrm{T}} V_{\mathrm{C}j} b_j^{\mathrm{T}}\right),$$

because both factors in the Kronecker product are scalars. Finally, $\mathrm{Var}\,(C) = \sum_{j=1}^{b} \mathrm{Cov}\,\left(C_j\right)$ because counts in different strata are mutually independent. The CMH correlation statistic S_{C} is now fully specified.

The CLT assures the asymptotic normality of C, so as the total sample size $n_{\bullet\bullet\bullet} = n_{\bullet\bullet 1} + \cdots + n_{\bullet\bullet b}$ approaches infinity, S_{C} has asymptotic distribution χ_1^2. Again see Landis et al. (1978). Alternatively, C is a scalar random variable with mean zero and is asymptotically univariate normal by the CLT. Thus $\frac{C^2}{\mathrm{Var}(C)} = S_{\mathrm{C}}$ is asymptotically distributed as χ_1^2.

2.5.2 An Alternative Presentation of the CMH C Test

Using the definitions previously given for V_{Tj} and V_{Cj}, we have

$$n_{\bullet\bullet j}a_j^T V_{Tj}a_j = \sum_{i=1}^{t} a_{ij}^2 n_{i\bullet j} - \frac{\left(\sum_{i=1}^{t} a_{ij}n_{i\bullet j}\right)^2}{n_{\bullet\bullet j}}$$

and

$$n_{\bullet\bullet j}b_j^T V_{Cj}b_j = \sum_{h=1}^{c} b_{hj}^2 n_{\bullet hj} - \frac{\left(\sum_{h=1}^{c} b_{hj}n_{\bullet hj}\right)^2}{n_{\bullet\bullet j}}.$$

On stratum j now define

$$S_{XXj} = \sum_{i=1}^{t} a_{ij}^2 n_{i\bullet j} - \frac{\left(\sum_{i=1}^{t} a_{ij}n_{i\bullet j}\right)^2}{n_{\bullet\bullet j}}$$

$$S_{XYj} = \sum_{i=1}^{t}\sum_{h=1}^{c} a_{ij}b_{hj}N_{ihj} - \frac{\left(\sum_{i=1}^{t} a_{ij}n_{i\bullet j}\right)\left(\sum_{h=1}^{c} b_{hj}n_{\bullet hj}\right)}{n_{\bullet\bullet j}} \qquad (2.1)$$

$$S_{YYj} = \sum_{h=1}^{c} b_{hj}^2 n_{\bullet hj} - \frac{\left(\sum_{h=1}^{c} b_{hj}n_{\bullet hj}\right)^2}{n_{\bullet\bullet j}}.$$

With appropriate divisors these quantities give unbiased estimators of the stratum variances and covariances. These expressions are familiar in formulae for regression coefficients.

Using the definitions in (2.1),

$$C_j = S_{XYj}, \quad n_{\bullet\bullet j}a_j^T V_{Tj}a_j = S_{XXj}, \quad \text{and} \quad n_{\bullet\bullet j}b_j^T V_{Cj}b_j = S_{YYj}.$$

Finally, note that since $C = \sum_{j=1}^{b} C_j$, $\text{Var}(C) = \sum_{j=1}^{b} \text{Var}(C_j)$. If we now write r_{Pj} for the Pearson correlation in the jth stratum, it follows that since $S_{XYj} = r_{Pj}\sqrt{S_{XXj}S_{YYj}}$,

$$S_C = \frac{C^2}{\text{Var}(C)} = \frac{\left(\sum_{j=1}^{b} S_{XYj}\right)^2}{\sum_{j=1}^{b}\frac{S_{XXj}S_{YYj}}{n_{\bullet\bullet j}-1}}$$

$$= \frac{\left(\sum_{j=1}^{b} r_{Pj}\sqrt{S_{XXj}S_{YYj}}\right)^2}{\sum_{j=1}^{b}\frac{S_{XXj}S_{YYj}}{n_{\bullet\bullet j}-1}}$$

$$= \left(\sum_{j=1}^{b} l_j r_{Pj}\right)^2,$$

in which $l_j = \sqrt{\dfrac{S_{XXj}S_{YYj}}{\sum_{j=1}^{b}\left(\frac{S_{XXj}S_{YYj}}{n_{\bullet\bullet j}-1}\right)}}$: S_C is the square of a linear combination of the

Pearson correlations in each stratum. This formula demonstrates how the Pearson correlations in each stratum contribute to the overall statistic.

The use of S_{XXj}, S_{XYj}, and S_{YYj} will now be demonstrated.

2.5.3 Examples

2.5.3.1 Homosexual Marriage Data

These data were considered in Section 1.4 and briefly in Section 2.1. Noting that stratum 1 is college and stratum 2 is school, we find $S_{XX1} = 42.63$, $S_{XY1} = -23.89$, $S_{YY1} = 51.67$, $r_{P1} = -0.5090$ and the CMH C statistic for school takes the value 18.66 with χ_1^2 p-value 0.0000. Similarly $S_{XX2} = 39.73$, $S_{XY2} = -9$, and $S_{YY2} = 50.00$, $r_{P2} = -0.2019$ and the CMH C statistic for college takes the value 2.4055 with χ_1^2 p-value 0.1209. From these the value of the CMH C statistic and its χ^2 p-value are confirmed: it was previously noted that S_C takes the value 16.8328 with χ_1^2 p-value 0.0000. Clearly there is a non-significant Pearson correlation for schools and a highly significant Pearson correlation for college. The latter dominates the former, so that overall there is strong evidence of a correlation effect: as religion becomes increasingly liberal there is greater agreement with the proposition that homosexuals should be able to marry. This is due mainly to the stratum college.

```
# Setting the treatment and response scores
a_ij = b_hj = matrix(rep(1:3,2), ncol = 2)
CMH(treatment = religion, response = opinion,
        strata = education, a_ij = a_ij, b_hj = b_hj,
        test_OPA = F, test_GA = F, test_MS = F)


##
##          Cochran Mantel Haenszel Tests
##
##                     S df    p-value
## Correlation 16.83   1 4.082e-05
##
##
## Correlations and statistics by strata:
##                 S_C   p-value       r_P  S_XX    S_XY  S_YY
## school        2.406 1.209e-01   -0.2019 39.73   -9.00 50.00
## college      18.660 1.566e-05   -0.5090 42.63  -23.89 51.67
```

Three special cases will be considered:

1. the data consists of one stratum only (as in the completely randomised design);

2. the treatment scores are independent of the strata: $a_{ij} = a_i$ for all i and j; and
3. the RBD.

In the first case the CMH correlation statistic simplifies to

$$S_C = \frac{(n-1)S_{XY}^2}{S_{XX}S_{YY}} = (n-1)\,r_P^2,$$

in which r_P is the Pearson correlation coefficient and there are n bivariate observations. This is well known. See, for example, Davis (2002, p. 253).

2.5.3.2 Whiskey Data

O'Mahony (1986, p. 363) originally gave the data in Table 2.3, which were analysed in Rayner and Best (2001). They use mid-rank scores and find the Spearman correlation, which takes the value -0.73. In testing if this is zero against two-sided alternatives, they give a Monte Carlo p-value of 0.09 and an asymptotic p-value of 0.04.

Using scores 1, 2, and 3 for grade and 1, 5, and 7 for years of maturity, we find $S_{XX} = 43.5$, $S_{XY} = -12$, and $S_{YY} = 6$. It follows that the Pearson correlation is -0.7428 and the CMH C statistic takes the value 3.862 with χ_1^2 p-value 0.0494. There is some evidence that as maturity increases, the grade of the whiskey improves. As χ_1^2 is an asymptotic distribution and there are only eight observations here it is highly desirable to calculate a resampling p-value. Using an observed CMH C test statistic of 3.862, permuting the treatment labels and calculating new statistics 1,000,000 times resulted in a p-value of 0.0747.

In the second special case S_{XX} is constant over strata. This gives a slight simplification of the S_C formula. If the data come from a RBD a considerable simplification is possible if the same treatment and response scores are used on each stratum or block. This will now be developed, with second and third special cases being demonstrated in the *Jams Data* in the following text after that development.

Table 2.3 Cross classification of age and whiskey.

Years of maturing	Grade			Total
	First	Second	Third	
One	0	0	2	2
Five	1	1	1	3
Seven	2	1	0	3
Total	3	2	3	8

```
CMH(treatment = maturity, response = grade, a_ij = c(1,5,7),
    b_hj = 1:3, test_OPA = F, test_GA = F, test_MS = F)

##
##          Cochran Mantel Haenszel Tests
##
##              S df p-value
## Correlation 3.862  1 0.04939
##
##
## Correlation and statistics:
##       r_P     S_XX     S_XY    S_YY
##   -0.7428  43.5000 -12.0000  6.0000
```

2.5.4 Derivation of the CMH C Test Statistic for the RBD with the Same Treatment Scores in Every Stratum

Suppose that the same treatment and response scores are used on each stratum or block. Then $a_{ij} = a_i$, say, for all i and j, and $b_{hj} = b_h$, say, for all h and j. Now $n_{i \bullet j} = 1$ for all i and j for the RBD, and as a consequence $n_{\bullet \bullet j} = t$ for all j. Thus

$$S_{XXj} = \sum_{i=1}^{t} a_{ij}^2 n_{i \bullet j} - \frac{\left(\sum_{i=1}^{t} a_{ij} n_{i \bullet j} \right)^2}{n_{\bullet \bullet j}} = \sum_{i=1}^{t} a_i^2 - \frac{\left(\sum_{i=1}^{t} a_i \right)^2}{t},$$

which is independent of j, so we may write $S_{XXj} = S_{XX}$. It is *not* true that S_{YYj} is independent of j. Next we focus on S_{YYj}. Put $x_{ij} = \sum_{h=1}^{c} b_{hj} N_{ihj}$. Now the N_{ihj} are all zero or one: treatment i on block j occupies one category h and is otherwise absent. In particular, for treatment i, over the c categories only once is N_{ihj} non-zero, and then it is one. Thus $x_{ij}^2 = \sum_{h=1}^{c} b_{hj}^2 N_{ihj}$. Consider the data set $\{x_{ij}\}$. If these are analysed as a *one-way ANOVA* or *completely randomised design* with *blocks as treatments* then the *error sum of squares* is

$$\sum_{i=1}^{t} \sum_{j=1}^{b} \left(x_{ij} - \bar{x}_{\bullet j} \right)^2 = \sum_{i=1}^{t} \sum_{j=1}^{b} x_{ij}^2 - t \sum_{j=1}^{b} \bar{x}_{\bullet j}^2 = \sum_{j=1}^{b} \left(\sum_{i=1}^{t} x_{ij}^2 - \frac{x_{\bullet j}^2}{t} \right)$$

$$= \sum_{j=1}^{b} \left[\sum_{i=1}^{t} \left(\sum_{h=1}^{c} b_{hj}^2 N_{ihj} \right) - \frac{\left(b_{hj} n_{\bullet hj} \right)^2}{t} \right]$$

$$= \sum_{j=1}^{b} \left[\sum_{h=1}^{c} b_{hj}^2 n_{\bullet hj} - \frac{\left(\sum_{h=1}^{c} b_{hj} n_{\bullet hj} \right)^2}{t} \right].$$

Finally, we focus on S_{XYj}. Write $\bar{a} = \sum_{i=1}^{t} \frac{a_i}{t}$. Then

$$\sum_{j=1}^{b} C_j = \sum_{i=1}^{t} \sum_{h=1}^{c} \sum_{j=1}^{b} a_i b_{hj} \left(N_{ihj} - \mathrm{E}\left[N_{ihj} \right] \right)$$

$$= \sum_{i=1}^{t} \sum_{h=1}^{c} \sum_{j=1}^{b} a_i b_{hj} \left(N_{ihj} - \frac{n_{\bullet hj}}{t} \right).$$

Now

$$\sum_{i=1}^{t} \sum_{h=1}^{c} \sum_{j=1}^{b} \left(a_i - \bar{a} \right) b_{hj} \left(N_{ihj} - \frac{n_{\bullet hj}}{t} \right)$$

$$= \sum_{i=1}^{t} \sum_{h=1}^{c} \sum_{j=1}^{b} \left(a_i - \bar{a} \right) \sum_{h=1}^{c} b_{hj} N_{ihj} - \sum_{i=1}^{t} \sum_{h=1}^{c} \sum_{j=1}^{b} \left(a_i - \bar{a} \right) \sum_{h=1}^{c} b_{hj} \frac{n_{\bullet hj}}{t}.$$

Since

$$\sum_{i=1}^{t} \sum_{h=1}^{c} \sum_{j=1}^{b} \left(a_i - \bar{a} \right) b_{hj} \frac{n_{\bullet hj}}{t} = \sum_{i=1}^{t} \left(a_i - \bar{a} \right) \sum_{h=1}^{c} b_{hj} \sum_{j=1}^{b} \frac{n_{\bullet hj}}{t}$$

$$= \left(\sum_{h=1}^{c} n_{\bullet h \bullet} b_{hj} \right) \sum_{i=1}^{t} \frac{a_i - \bar{a}}{t} = 0,$$

it follows that

$$C = \sum_{j=1}^{b} C_j = \sum_{i=1}^{t} \left(a_i - \bar{a} \right) \sum_{h=1}^{c} \sum_{j=1}^{b} b_{hj} N_{ihj}.$$

Note that $\sum_{h=1}^{c} \sum_{j=1}^{b} b_h N_{ihj}$ is the sum of the ith treatment scores over responses and blocks, and these are usually easy to calculate directly or are readily available in most packaged analyses. The $\left(a_i - \bar{a} \right)$ are the centred treatment scores. Thus

$$S_C = \frac{(t-1)\, C^2}{S_{XX} \sum_{j=1}^{b} S_{YYj}},$$

in which C is the sum of the products of the treatment sums and the centred treatment scores, S_{XX} is the sum of the squares of the centred treatment scores and $\sum_{j=1}^{b} S_{YYj}$ may be read from the output for a one-way ANOVA. It is the error sum of squares when blocks are taken to be treatments.

2.5.4.1 Jams Data Revisited

As there are eight strata, hand calculation is possible if a little tedious. Because S_{XX} is constant over strata it is not too much extra work to calculate S_{XY}, S_{YY}, the Pearson correlation, the CMH correlation statistic and its p-value on each stratum. A summary of the calculations is given in Table 2.4.

Table 2.4 Analysis of Jams data.

	Stratum							
	1	**2**	**3**	**4**	**5**	**6**	**7**	**8**
S_{XY}	0	0	0	1	0	2	2	0
S_{YY}	0.6667	0.6667	0.6667	4.6667	2.6667	2.6667	4.6667	6
r_P	0	0	0	0.3273	0	0.8660	0.6547	0
S_C	0	0	0	0.2143	0	1.5	0.8571	0
p-value	1	1	1	0.6434	1	0.2207	0.3545	1

For these data $a_i = 1, 2,$ and 3, so $S_{XX} = 2$. From a one-way ANOVA with judges as treatments, $\sum_{j=1}^{8} S_{YYj} = \frac{68}{3}$. This can be confirmed by summing across the S_{YY} row in Table 2.4. By summing the columns in Table 2.4 the treatment score sums are found to be 18, 30, and 23, and since the centred treatment scores are -1, 0, and 1, $C = -18 + 0 + 23 = 5$. Substituting gives $S_C = 2 \times \frac{25}{2 \times \frac{68}{3}} = \frac{75}{68} = 1.1029$ with p-value 0.2936, as given in Section 2.1. There is no significant correlation effect, which would, if present, indicate that as we pass from jam A to B to C there is increasing (or decreasing) sweetness. It could be that overall there is no correlation effect with the contrary being the case in a minority of strata. That is not the case here; no stratum shows any evidence of even a slight correlation effect. Here this is hardly surprising; with only three observations in each stratum there can be little power in testing for a correlation.

```
# Setting the treatment and response scores
a_ij = matrix(rep(1:3,8), ncol = 8)
b_hj = matrix(rep(1:5,8), ncol = 8)
CMH(treatment = type, response = sweetness, strata = judge,
          a_ij = a_ij, b_hj = b_hj, test_OPA = F, test_GA = F,
          test_MS = F)

##
##          Cochran Mantel Haenszel Tests
##
##                    S df p-value
## Correlation 1.103   1  0.2936
##
##
## Correlations and statistics by strata:
##       S_C p-value    r_P S_XX S_XY    S_YY
## 1 0.0000  1.0000 0.0000    2    0 0.6667
## 2 0.0000  1.0000 0.0000    2    0 0.6667
## 3 0.0000  1.0000 0.0000    2    0 0.6667
```

```
## 4 0.2143  0.6434 0.3273    2    1 4.6670
## 5 0.0000  1.0000 0.0000    2    0 2.6670
## 6 1.5000  0.2207 0.8660    2    2 2.6670
## 7 0.8571  0.3545 0.6547    2    2 4.6670
## 8 0.0000  1.0000 0.0000    2    0 6.0000
```

If there is interest in the contributions to the correlation from individual strata (as in the Homosexual Marriage Data) this is a reasonable approach. However, if there is not, then for the RBD with the same treatment scores in each stratum this approach simplifies hand calculation.

2.5.5 The CMH C Test Statistic Is Not, in General, Location-Scale Invariant

It will subsequently be of interest to find S_C when the scores have been transformed. One obvious transformation to consider is standardisation. First put $p_{i \bullet j} = \frac{N_{i \bullet j}}{n_{\bullet \bullet j}}$ and $p_{\bullet h j} = \frac{N_{\bullet h j}}{n_{\bullet \bullet j}}$, and $c_{Xj} = \sum_{i=1}^{t} a_{ij} p_{i \bullet j}$ and $d_{Xj}^2 = \sum_{i=1}^{t} \left(a_{ij} - c_{Xj} \right)^2 p_{i \bullet j}$. Instead of the scores $\{ a_{ij} \}$ use the standardised scores $\left\{ \frac{a_{ij} - c_{Xj}}{d_{Xj}} \right\}$. These scores have mean zero since

$$\sum_{i=1}^{t} \frac{\left(a_{ij} - c_{Xj} \right) p_{i \bullet j}}{d_{Xj}} = \sum_{i=1}^{t} \frac{a_{ij} p_{i \bullet j}}{d_{Xj}} - c_{Xj} \sum_{i=1}^{t} \frac{p_{i \bullet j}}{d_{Xj}} = 0.$$

Similarly put $e_{Yj} = \sum_{i=1}^{t} b_{hj} p_{\bullet h j}$ and $f_{Yj}^2 = \sum_{h=1}^{c} \left(b_{\bullet h j} - e_{Yj} \right)^2 p_{\bullet h j}$ and instead of the scores $\{ b_{hj} \}$ use the standardised scores $\left\{ \frac{b_{hj} - e_{Yj}}{f_{Yj}} \right\}$. As with the other set of standardised scores, these scores have mean zero.

Now $S_{XXj} = n_{\bullet \bullet j} d_{Xj}^2$ and the corresponding quantity using the standardised scores is $\sum_{i=1}^{t} \left(\frac{a_{ij} - c_{Xj}}{d_{Xj}} \right)^2 n_{i \bullet j} = n_{\bullet \bullet j} d_{Xj}^2 = S_{UUj}$, say. Similarly $S_{YYj} = n_{\bullet \bullet j} f_{Yj}^2 = S_{VVj}$, say. To give the CMH C test statistic using the standardised scores we now need $S_{UVj} = \sum_{i=1}^{t} \frac{(a_{ij} - c_{Xj})(b_{\bullet h j} - e_{Yj}) p_{i \bullet j}}{d_{Xj} f_{Yj}} = \frac{S_{XYj}}{d_{Xj} f_{Yj}}$. Hence the CMH C test statistic using the original scores is

$$S_{CXY} = \frac{\left(\sum_{j=1}^{b} S_{XYj} \right)^2}{\sum_{j=1}^{b} \frac{S_{XXj} S_{YYj}}{n_{\bullet \bullet j} - 1}}$$

$$= \frac{\left(\sum_{j=1}^{b} d_{Xj} f_{Yj} S_{UVj} \right)^2}{\sum_{j=1}^{b} \frac{d_{Xj}^2 f_{Yj}^2 S_{UUj} S_{VVj}}{n_{\bullet \bullet j} - 1}}.$$

If the $\{d_{Xj}\}$ do not depend on j and similarly $\{f_{Yj}\}$ do not depend on j, that is, the scale factors are the same across strata, then the CMH C tests using the original scores and the standardised scores are the same. However, in general this cannot be expected.

Chapter 3 focuses on the completely randomised design, which is a special case of the CMH design. The Kruskal–Wallis test is shown to be a special case of the CMH MS statistic, and interesting results are shown to hold for the CMH tests for this design.

Bibliography

Cochran, W. G. (1954). Some methods for strengthening the common χ^2 tests. *Biometrics*, 10(4):417–451.

Davis, C. S. (2002). *Statistical methods for the analysis of repeated measurements*. Technical report, Springer.

Irwin, J. O. (1949). A note on the subdividsion of χ^2 into components. *Biometrika*, 36(1/2):130–134.

Kuritz, S. J., Landis, J. R., and Koch, G. G. (1988). A general overview of Mantel-Haenszel methods: applications and recent developments. *Annual Review of Public Health*, 9(1):123–160.

Lancaster, H. O. (1949). The derivation and partition of $\chi 2$ in certain discrete distributions. *Biometrika*, 36(1-2):117–129.

Landis, J. R., Cooper, M. M., Kennedy, T., and Koch, G. G. (1979). A computer program for testing average partial association in three-way contingency tables (PARCAT). *Computer Programs in Biomedicine*, 9(3):223–246.

Landis, J. R., Heyman, E. R., and Koch, G. G. (1978). Average partial association in three-way contingency tables: a review and discussion of alternative tests. *International Statistical Review/Revue Internationale de Statistique*, 46(3):237–254.

Mantel, N. and Haenszel, W. (1959). Statistical aspects of the analysis of data from retrospective studies of disease. *Journal of the National Cancer Institute*, 22(4):719–748.

O'Mahony, M. (1986). *Sensory Evaluation of Food: Statistical Methods and Procedures*. New York: Marcel Dekker.

Rayner, J. C. W. and Best, D. J. (2001). *A Contingency Table Approach to Nonparametric Testing*. Boca Raton, FL: Chapman & Hall/CRC.

Rayner, J. C. W. and Best, D. J. (2017). Unconditional analogues of Cochran-Mantel-Haenszel tests. *Australian & New Zealand Journal of Statistics*, 59(4):485–494.

3

The Completely Randomised Design

3.1 Introduction

The completely randomised design (CRD) is often the first experimental design met by students and users of statistics. It has wide applicability and is easily understood. When the data are consistent with normality analysis can proceed using the ANOVA F test. When that is not the case the data may be ranked, and a Kruskal–Wallis test is appropriate. However, in some scenarios the responses are categorical, and then Cochran–Mantel–Haenszel (CMH) methods may be applied, for the CRD is consistent with what we have called the CMH design.

In this chapter, we first give the parametric model. It will be needed for subsequent derivations, and even when the assumptions are not obviously met, the well-known robustness of the ANOVA models suggests the resulting conclusions may nevertheless be valid. Ranking methods lead to the Kruskal–Wallis tests, for both when the data are distinct, and when there are ties.

We show how to adjust the Kruskal–Wallis to deal with ties. This adjusted Kruskal–Wallis test statistic is simply related to the ANOVA F test statistic when the data are the ranks. This shows that the test statistics represent different forms of the same test. Hence, both are testing the same null hypothesis: equality of treatment rank means.

Attention will be focused on when there are ties and mid-rank scores are used, since this is perhaps the most common method for dealing with ties.

The CMH tests are developed for the particular scenario of the CRD, for then the CMH test statistics take forms simpler than the most general, described in the previous chapter. We show that the Kruskal–Wallis tests, even for tied data, are CMH tests. A simulation study shows that when using mid-ranks the F approximation is superior to the χ^2 approximation close to the nominal significance level. Finally, we show that the CMH tests

An Introduction to Cochran–Mantel–Haenszel Testing and Nonparametric ANOVA,
First Edition. J.C.W. Rayner and G. C. Livingston Jr.
© 2023 John Wiley & Sons Ltd. Published 2023 by John Wiley & Sons Ltd.

are also Wald-type tests, which mean the optimality properties of these tests are conferred upon these CMH tests.

3.2 The Design and Parametric Model

In the CRD, there is one *factor* or effect of interest. It is observed at different *levels*. The question of interest is whether or not the levels are different and so the null hypothesis is that the mean levels are equal. The initial applications of the analyses associated with this design were in agricultural experiments. Treatments were allocated *randomly* to the plots used. Hence, the name 'completely randomised' design.

In the one-factor ANOVA, we observe random variables Y_{ij}, $j = 1, \ldots, n_i$, $i = 1, \ldots, t$. There are $n = \sum_{i=1}^{t} n_i$ observations in all. The parametric model is that the Y_{ij} are $IN\left(\mu_i, \sigma^2\right)$: all observations are mutually independent, normally distributed and have a constant variance. The quantity $CF = \frac{\left(\sum_{i=1}^{t} \sum_{j=1}^{n_i} Y_{ij}\right)^2}{n}$ is called a correction factor. Define the total, factor, and error sums of squares by

$$SS_{\text{Total}} = \sum_{i=1}^{t} \sum_{j=1}^{n_i} Y_{ij}^2 - CF,$$

$$SS_{\text{Treat}} = \sum_{i=1}^{t} \frac{Y_{i\bullet}^2}{n_i} - CF,$$

$$SS_{\text{Error}} = SS_{\text{Total}} - SS_{\text{Treat}},$$

respectively. As before, a dot replacing a subscript means summation over that subscript. These sums of squares are associated with $n-1$, $t-1$, and $n-t$ degrees of freedom, respectively. The corresponding mean squares are the sums of squares divided by their degrees of freedom: $MS_{\text{Treat}} = \frac{SS_{\text{Treat}}}{t-1}$ and $MS_{\text{Error}} = \frac{SS_{\text{Error}}}{n-t}$. The ANOVA F test statistic is $F = \frac{MS_{\text{Treat}}}{MS_{\text{Error}}}$ and has sampling distribution $F_{t-1,n-t}$. The null hypothesis under this model, that the treatment means are all equal, $\mu_1 = \cdots = \mu_t$, is tested against at least one pair of population means being not equal.

In the next section, we look at rank-based nonparametric competitors for the parametric F test: the Kruskal–Wallis tests.

3.3 The Kruskal–Wallis Tests

If the assumptions for the one factor ANOVA (normality, constant variance) are not satisfied, an alternative test that doesn't make such assumptions is needed to test for treatment effects. The Kruskal–Wallis test is based on ranks and is a generalisation of the Wilcoxon/Mann–Whitney test. Conventional wisdom is that the null hypothesis for the Kruskal–Wallis test is equality of the treatment distributions, although it is known that the test is sensitive to location differences. We will challenge that view later in this chapter.

Suppose we have distinct (untied) observations. All n observations are combined, ordered, and ranked. For $i = 1, \ldots, t$ the sum of the ranks for treatment i, R_i, is calculated. The Kruskal–Wallis test statistic KW is given by

$$KW = -3(n+1) + \frac{12}{n(n+1)} \sum_{i=1}^{t} \frac{R_i^2}{n_i}.$$

If ties occur, a correction to KW is required. See, for example, Thas et al. (2012). This correction is based on the r_{ij}, the overall rank of y_{ij}. The assignment of ranks is made so that the rank sum is $\frac{n(n+1)}{2}$, the same as the rank sum for untied data. The adjusted Kruskal–Wallis test statistic is

$$KW_A = \frac{(n-1)\left(\sum_{i=1}^{t} \frac{R_i^2}{n_i} - \frac{n(n+1)^2}{4}\right)}{\sum_{i=1}^{t} \sum_{j=1}^{n_i} r_{ij}^2 - \frac{n(n+1)^2}{4}}.$$

The basis for this correction is to replace the variance of the untied ranks, $\frac{(n-1)(n+1)}{12}$, by the variance of the tied ranks, $\sum_{i=1}^{t} \sum_{j=1}^{n_i} \frac{r_{ij}^2}{n} - \left(\frac{n+1}{2}\right)^2$. We emphasise that this form does *not* depend on the use of mid-ranks in the treatment of the ties.

If there are no ties, then KW_A reduces to KW. To show this, the following identities are required:

$$1 + \cdots + n = \frac{n(n+1)}{2}$$

$$1^2 + \cdots + n^2 = \frac{n(n+1)(2n+1)}{6} \qquad (3.1)$$

$$\frac{n(n+1)(2n+1)}{6} - n\left(\frac{n+1}{2}\right)^2 = \frac{(n-1)(n+1)}{12}.$$

If there are no ties, then the ranks are $1, 2, \ldots, n$, and the denominator in KW_A, by the second and third of the given identities, is $\frac{(n-1)n(n+1)}{12}$, then KW_A reduces to KW.

An alternative correction is now derived for when there are ties and mid-ranks are used. To distinguish this test from the completely general adjusted Kruskal–Wallis test, we will henceforth use the notation KW_M for the test statistic when mid-ranks are used. We now derive KW_M.

First, we need to point out that while one rationale for using mid-ranks is that doing so preserves the sum of the ranks, their use reduces the sum of the squares of the ranks and therefore their variance, as the following lemma demonstrates. See also Gibbons and Chakraborti (2021).

3.3.1 Mid-Rank Lemma

Suppose a group of t observations, which would otherwise have occupied the ranks $h + 1, h + 2, \ldots, h + t$, are tied. Each rank is replaced by the mean of these ranks, $h + \frac{t+1}{2}$. This reduced the sum of squares by $\frac{t^3 - t}{12}$.

Proof

The sum of the squares of the ranks $h + 1, h + 2, \ldots, h + t$ is

$$(h + 1)^2 + \cdots + (h + t - 1)^2 + (h + t)^2$$

$$= th^2 + 2h \sum_{j=1}^{t} j + \sum_{j=1}^{t} j^2 = th^2 + t(t + 1)h + \frac{t(t + 1)(2t + 1)}{6}$$

using the well-known formulae for the sum and the sum of the squares of the first t integers. If mid-ranks are assigned, the sum of these squares is

$$t\left(h + \frac{t + 1}{2}\right)^2 = t\left(h^2 + (t + 1)h + \frac{t(t + 1)^2}{4}\right)$$

and the difference is

$$\frac{t(t + 1)(2t + 1)}{6} - \frac{t(t + 1)^2}{4} = \frac{t^3 - t}{12}.$$

There may be multiple groups and suppose the gth group out of G groups is of size t_g. In any group, changes to the sum of squares due to replacing untied ranks by mid-ranks do not affect changes in any other group. Hence, such changes are additive. It follows that when passing from untied to tied data using mid-ranks, the sum of squares is reduced by the aggregation of the corrections $\frac{t_g^3 - t_g}{12}$ for all groups of tied observations. Thus, $1^2 + \cdots + n^2$ becomes $1^2 + \cdots + n^2 - \sum_g \frac{t_g^3 - t_g}{12}$.

In KW_A, the denominator $\sum_{i=1}^{t} \sum_{j=1}^{n_i} r_{ij}^2 - \frac{n(n+1)^2}{4}$ becomes

$$1^2 + \cdots + n^2 - \sum_g \frac{t_g^3 - t_g}{12} - n\left(\frac{n + 1}{2}\right)^2$$

$$= \frac{(n - 1)n(n + 1)}{12} - \sum_g \frac{t_g^3 - t_g}{12} = \frac{C_{\mathrm{CRD}}(n - 1)n(n + 1)}{12}$$

provided, we define

$$C_{\mathrm{CRD}} = 1 - \sum_{g=1}^{G} \frac{t_g^3 - t_g}{(n - 1)n(n + 1)}.$$

Thus,

$$C_{CRD} KW_M = \frac{12(n-1)\left(\sum_{i=1}^{t} \frac{R_i^2}{n_i} - \frac{n(n+1)^2}{4}\right)}{(n-1)n(n+1)}$$

$$= \frac{12\sum_{i=1}^{t} \frac{R_i^2}{n_i}}{n(n+1)} - 3(n+1)$$

$$= KW$$

and $KW_M = \frac{KW}{C_{CRD}}$. We use the notation C_{CRD} because for other designs there will be analogous but different constants.

3.3.2 Whiskey Data Revisited

See Section 2.5.3. Using mid-ranks for the tied data, the rank sums for the treatments, one, five, and seven years of maturing, are 14, 13.5, and 8.5, respectively. Thus,

$$KW = \frac{12}{8 \times 9}\left[\frac{14^2}{2} + \frac{13.5^2}{3} + \frac{8.5^2}{3}\right] - 3 \times 9$$

$$= \frac{98 + 60.75 + 24.0833}{6} - 27 = 3.4722.$$

There are three groups of ties, of sizes 3, 2, and 3. For $t_g = 2$ and 3, $t_g^3 - t_g = 6$ and 24, respectively, so that $C_{CRD} = 1 - \frac{24+6+24}{7 \times 8 \times 9} = \frac{25}{28}$ and $KW_M = 3.8889$. The χ_2^2 p-values for KW and KW_M are 0.1762 and 0.1431, respectively. Direct calculation gives $\sum_{i=1}^{t} \sum_{j=1}^{n_i} r_{ij}^2 = 199.5$ and $KW_A = 3.8889$.

```
KW(treatment = maturity, response = grade)

##
##          Kruskal--Wallis Rank Sum Test
##
## data:  grade and maturity
## Adjusting for ties:
## KW_A = 3.888889, df = 2, p-value = 0.1430667
##
## No adjustment for ties:
## KW = 3.472222, df = 2, p-value = 0.1762043
```

3.4 Relating the Kruskal–Wallis and ANOVA F Tests

When there are no ties, the relationship between the Kruskal–Wallis and ANOVA F tests is well known. For example, Rayner (2016) shows that in this case if F is the ANOVA F test statistic, then

$$F = \frac{(n-t)KW}{(t-1)(n-1-KW)}.$$

The $F_{t-1,n-1-t}$ distribution of F can then be used to calculate p-values. In this section, we show that the relationship generalises to between F and the Kruskal–Wallis statistic adjusted for ties, KW_A. Thus, even when there are ties the Kruskal–Wallis test statistic and the corresponding p-value can be calculated by using ANOVA software and the relationship.

The observations are the $r_{ij}, j = 1, \ldots, n_i, i = 1, \ldots, t$. There are $n = \sum_{i=1}^{t} n_i$ observations in all and the sum of the observations is $r_{\bullet\bullet} = \frac{n(n+1)}{2}$. The correction factor is $CF = \frac{\left(\sum_{i=1}^{t} \sum_{j=1}^{n_i} r_{ij}\right)^2}{n} = \frac{r_{\bullet\bullet}^2}{n} = \frac{n(n+1)^2}{4}$. Thus,

- the total sum of squares is $SS_{\text{Total}} = \sum_{i=1}^{t} \sum_{j=1}^{n_i} r_{ij}^2 - CF$
- the treatment sum of squares is $SS_{\text{Treat}} = \sum_{i=1}^{t} \frac{R_i^2}{n_i} - CF$
- the error sums of squares is $SS_{\text{Error}} = SS_{\text{Tot}} - SS_F = \sum_{i=1}^{t} \sum_{j=1}^{n_i} r_{ij}^2 - \sum_{i=1}^{t} \sum_{j=1}^{n_i} \frac{R_i^2}{n_i}$.

As in the previous section, Section 3.3, the adjusted Kruskal–Wallis test statistic is

$$KW_A = \frac{(n-1)\left(\sum_{i=1}^{t} \frac{R_i^2}{n_i} - CF\right)}{\sum_{i=1}^{t} \sum_{j=1}^{n_i} r_{ij}^2 - CF}.$$

Thus

$$SS_{\text{Treat}} = \sum_{i=1}^{t} \frac{R_i^2}{n_i} - CF = \frac{KW_A \left(\sum_{i=1}^{t} \sum_{j=1}^{n_i} r_{ij}^2 - CF\right)}{n-1},$$

$$SS_{\text{Error}} = \sum_{i=1}^{t} \sum_{j=1}^{n_i} r_{ij}^2 - \sum_{i=1}^{t} \frac{R_i^2}{n_i}$$

$$= \sum_{i=1}^{t} \sum_{j=1}^{n_i} r_{ij}^2 - CF - \left(\sum_{i=1}^{t} \frac{R_i^2}{n_i} - CF\right)$$

$$= \sum_{i=1}^{t} \sum_{j=1}^{n_i} r_{ij}^2 - CF - \frac{KW_A \left(\sum_{i=1}^{t} \sum_{j=1}^{n_i} r_{ij}^2 - CF\right)}{n-1}.$$

The F test statistic is thus

$$F = \frac{(n-t) SS_{\text{Treat}}}{(t-1) SS_{\text{Error}}} = \frac{(n-t)\left(\sum_{i=1}^{t} \frac{R_i^2}{n_i} - CF\right)}{(t-1)\left(\sum_{i=1}^{t} \sum_{j=1}^{n_i} r_{ij}^2 - CF\right)}$$

$$= \frac{(n-t) KW_A \left(\sum_{i=1}^{t} \sum_{j=1}^{n_i} r_{ij}^2 - CF\right)(n-1)}{(t-1)\left[\left(\sum_{i=1}^{t} \sum_{j=1}^{n_i} r_{ij}^2 - CF\right) - \frac{KW_A \left(\sum_{i=1}^{t} \sum_{j=1}^{n_i} r_{ij}^2 - CF\right)}{n-1}\right]}$$

$$= \frac{(n-t) KW_A}{(t-1)\left(n-1-KW_A\right)}.$$

This is a purely algebraic relationship, requiring no model. It means the F test is merely a different form of the adjusted Kruskal–Wallis test. If the sampling distributions were known exactly each would give exactly the same inference as the other. However, the sampling distribution of the adjusted Kruskal–Wallis test is only known asymptotically, while that of the ANOVA F test only known approximately, since the data are ranks and not distributed normally.

The relationship means that the two tests coincide, and hence are testing the same null hypothesis against the same alternative. Recall that the Kruskal–Wallis purportedly tests for equality of the treatment distributions with the caveat that it is known to be sensitive to location differences. However, the ANOVA F test tests for equality of treatment mean ranks, and hence, so does the Kruskal–Wallis, whether or not it is adjusted for ties.

3.5 The CMH Tests for the CRD

For the CRD, there is only one stratum, and the CMH overall partial association (OPA) and general association (GA) tests coincide. From Section 2.3, the usual CMH OPA test statistic S_{OPA} is the sum over strata of weighted Pearson X^2 statistics, the weights being $\frac{n_{\bullet\bullet j}-1}{n_{\bullet\bullet j}}$. As there is only one stratum, there are $n_{i\bullet\bullet} = n_i$ observations on the ith treatment, and $n = n_1 + \cdots + n_t$ observations in all. Thus,

$$
\begin{aligned}
W_{OPA} &= W_{GA} \\
&= \frac{n-1}{n}\sum_{i=1}^{t}\sum_{h=1}^{c}\frac{\left(N_{ih\bullet} - \frac{n_i n_{\bullet h\bullet}}{n}\right)^2}{\frac{n_i n_{\bullet h\bullet}}{n}} \\
&= \frac{n-1}{n}X_P^2,
\end{aligned}
$$

in which X_P^2 is the usual Pearson test statistic. Previously, we used the notation S_{MS} for the CMH tests in general; here they will be denoted by W_{MS} for the particular expressions for the CRD.

Now, suppose a score b_h is assigned to the hth response. The score sum for treatment i is then $\sum_{h=1}^{c}\sum_{j=1}^{b}b_h N_{ihj} = \sum_{h=1}^{c}b_h N_{ih\bullet} = M_i$. Using the extended hypergeometric distribution M_i has expectation $E\left[M_i\right] = n_i\sum_{h=1}^{c}b_h\frac{n_{\bullet h\bullet}}{n}$ under the null hypothesis of no mean effects. Write

$$
V_i = M_i - E\left[M_i\right] = \sum_{h=1}^{c}b_h N_{ih\bullet} - n_i\sum_{h=1}^{c}\frac{b_h n_{\bullet h\bullet}}{n},
$$

$$
(n-1)S^2 = n\sum_{h=1}^{c}b_h^2 n_{\bullet h\bullet} - \left(\sum_{h=1}^{c}b_h n_{\bullet h\bullet}\right)^2,
$$

$$
V = \{V_i\}, \text{ and } D = \text{diag}\left(\sqrt{\frac{n_i}{n}}\right).
$$

Then it will be shown in Section 3.9.2 that

$$W_{MS} = \frac{V^T D^{-2} V}{S^2} = \frac{n}{S^2} \sum_{i=1}^{t} \frac{V_i}{n_i}.$$

Finally, suppose a score a_i is assigned to the ith treatment while, as for the mean scores (MS) statistic, a score b_h is assigned to the hth response. Put

$$S_{XX} = \sum_{i=1}^{t} a_i^2 n_i - \frac{\left(\sum_{i=1}^{t} a_i n_i\right)^2}{n},$$

$$S_{XY} = \sum_{i=1}^{t} \sum_{h=1}^{c} a_i b_h N_{ih\bullet} - \frac{\left(\sum_{i=1}^{t} a_i n_i\right)\left(\sum_{h=1}^{c} b_h n_{\bullet h\bullet}\right)}{n},$$

$$S_{YY} = \sum_{h=1}^{c} b_h^2 n_{\bullet h\bullet} - \frac{\left(\sum_{h=1}^{c} b_h n_{\bullet h\bullet}\right)^2}{n}.$$

Then as in Section 2.5.2

$$W_C = \frac{(n-1) S_{XY}^2}{S_{XX} S_{YY}} = (n-1) r_P^2,$$

in which r_P is the Pearson correlation coefficient. See, for example, Davis (2002, p. 253).

3.5.1 Whiskey Data Revisited

Routine calculations find that $X_P^2 = -n + \sum \frac{observed^2}{expected} = \frac{16}{3}$, from which $\frac{n-1}{n} X_P^2 = \frac{14}{3} = 4.667 = W_{GA}$. The χ_4^2 p-value is 0.3232. For these data at the 0.05 level of significance, the GA test finds no evidence of an association between years of maturing and grade of whiskey.

Using scores 1, 2, and 3 for grade, hand calculations give $\sum_{h=1}^{c} b_h N_{ih\bullet} = 16$, $\sum_{h=1}^{c} b_h^2 n_{\bullet h\bullet} = 44$, $S^2 = \frac{48}{7}$, $V_1 = -2$, $V_2 = 0$, $V_3 = 2$ and $W_{MS} = \frac{35}{9}$. This gives a χ_2^2 p-value of 0.1431, the same as found from applying KW_A. At the usual levels of significance, there is no evidence that the whiskey MS differ.

Using scores 1, 2, and 3 for grade and 1, 5, and 7 for years of maturity, in Section 2.5.2, using the formula given earlier for W_C, we found that the CMH C statistic takes the value 3.8621 with χ_1^2 p-value just less than 0.05 (0.0494). There is some evidence that as maturity increases so does the grade of the whiskey.

The tests based on W_{GA}, W_{MS}, and W_C seek alternatives in parameter spaces of dimensions 4, 2, and 1, respectively, the degrees of freedom associated with the χ^2 sampling distributions. Thus, the tests based on W_{GA}, W_{MS} are less focused than the CMH C test, which is one reason they fail to detect an effect.

In Section 3.9, models will be given from which Wald test statistics can be derived. These will be shown to be the test statistics introduced here.

3.6 The KW Tests Are CMH MS Tests

In this section, we show that subject to a very mild condition the Kruskal–Wallis tests adjusted for tied data are CMH MS tests.

We begin by making the notations compatible. If there are $c < n$ distinct ranks assume that there are c categorises and assign to these categories distinct scores the ranks, b_h, say, with $b_1 < b_2 < \cdots < b_c$. Define $N_{ihj} = 1$ if the jth replicate of treatment i is given rank b_h and zero otherwise. Then $b_h N_{ihj} = r_{ij}$, the rank given to the jth replicate of treatment i.

We assume that $\sum_{h=1}^{c} b_h N_{\bullet h \bullet} = \sum_{i=1}^{t} \sum_{j=1}^{b} r_{ij} = \frac{n(n+1)}{2}$: that the assignment of ranks is such that the rank sum is the same as it would be for distinct ranks.

From Section 3.5 (and Section 3.9.2), we have

$$V_i = \sum_{h=1}^{c} b_h N_{ih\bullet} - n_i \sum_{h=1}^{c} \frac{b_h n_{\bullet h \bullet}}{n},$$

$$(n-1)S^2 = n \sum_{h=1}^{c} b_h^2 n_{\bullet h \bullet} - \left(\sum_{h=1}^{c} b_h n_{\bullet h \bullet} \right)^2, \text{ and}$$

$$W_{MS} = \frac{n}{S^2} \sum_{i=1}^{t} \frac{V_i^2}{n_i}.$$

First, $\sum_{h=1}^{c} b_h N_{ih\bullet} = \sum_{j=1}^{b} r_{ij} = R_i$, the rank sum for treatment i and $\sum_{h=1}^{c} b_h n_{\bullet h \bullet} = \frac{n(n+1)}{2}$ by the assumption earlier. Thus, $V_i = R_i - \frac{n_i(n+1)}{2}$ and

$$\sum_{i=1}^{t} \frac{V_i^2}{n_i} = \sum_{i=1}^{t} \frac{\left(R_i - \frac{n_i(n+1)}{2} \right)^2}{n_i} = \sum_{i=1}^{t} \frac{R_i^2}{n_i} - \frac{n(n+1)^2}{4}.$$

Second, $\sum_{h=1}^{c} b_h^2 N_{ihj} = r_{ij}^2$ so that $\sum_{h=1}^{c} b_h^2 N_{\bullet h \bullet} = \sum_{i=1}^{t} \sum_{j=1}^{b} r_{ij}^2$. Thus

$$(n-1)S^2 = n \sum_{i=1}^{t} \sum_{j=1}^{b} r_{ij}^2 - \left[\frac{n(n+1)}{2} \right]^2$$

and

$$\frac{S^2}{n} = \frac{\sum_{i=1}^{t} \sum_{j=1}^{b} r_{ij}^2 - \frac{n(n+1)^2}{4}}{n-1}.$$

Substituting in W_{MS} gives KW_A as defined in Section 3.3. Thus $W_{MS} = KW_A$.

3.7 Relating the CMH MS and ANOVA F Tests

We established in Section 3.4 that

$$F = \frac{(n-t)\,KW_A}{(t-1)\left(n-1-KW_A\right)}.$$

However, the KW and KW_A tests are particular cases of CMH MS tests; the latter may have scores other than the ranks, and so are more general. Although the derivation did depend strongly on the CMH MS scores, it is possible the relationship extends to the same relationship between the CMH MS statistic for the CRD and the ANOVA F statistic. Next, we show this is indeed the case. Since the CMH MS test is a location test, this reinforces the previous conclusion that so are these Kruskal–Wallis tests.

It is also of interest that the ANOVA F statistic calculated from mid-ranks is the *rank transform statistic* for this design. Rank transform statistics result in any design when the data are replaced by their ranks, either overall or within subgroups. The development here shows that the rank transform statistic calculated from mid-ranks for the ANOVA F statistic in the CRD is a CMH MS statistic.

3.7.1 The One-Way ANOVA F Test

Recall the ANOVA from Section 3.2, in which the data were $y_{ij}, j = 1, \ldots, n_i$ and $i = 1, \ldots, t$. Write T_i for $\sum_{j=1}^{b} y_{ij}$ and $T_{\bullet\bullet}$ for $\sum_{i=1}^{t} T_i$. If we put $CF = \frac{T_{\bullet\bullet}^2}{n}$, a correction factor, then the factor sum of squares, SS_{Treat}, is $\sum_{i=1}^{t} T_i^2 - CF$, the total sum of squares, SS_{Total}, is $\sum_{i=1}^{t} \sum_{j=1}^{b} y_{ij}^2 - CF$, and the error sum of squares, SS_{Error}, is $SS_{\text{Total}} - SS_{\text{Treat}}$. The ANOVA F statistic is $F = \frac{\frac{SS_{\text{Treat}}}{t-1}}{SS_{\text{Error}}/(n-t)}$.

3.7.2 W_{MS} in Terms of the ANOVA F

In the CMH setting, we have counts and scores. Suppose that for $i = 1, \ldots, t$ and $h = 1, \ldots, c$, $N_{ihj} = 1$ if the jth observation on the ith treatment is assessed to be in the hth category, and zero otherwise. As before, a score b_h is assigned to the hth response. We use $\sum_{h=1}^{c} b_h N_{ihj}$ as data for an ANOVA F test; that is, put $y_{ij} = \sum_{h=1}^{c} b_h N_{ihj}$. Now $T_i = \sum_{h=1}^{c} b_h N_{ih\bullet}$ and $T_{\bullet\bullet} = \sum_{h=1}^{c} b_h N_{\bullet h\bullet}$.

For the jth replicate of treatment i, there is only one value of h for which n_{ihj} is equal to 1; otherwise, $n_{ihj} = 0$. Thus, $y_{ij}^2 = \sum_{h=1}^{c} b_h^2 N_{ihj}$ and $\sum_{i=1}^{t} \sum_{j=1}^{b} y_{ij}^2 = \sum_{i=1}^{t} \sum_{h=1}^{c} \sum_{j=1}^{b} b_h^2 N_{ihj} = \sum_{h=1}^{c} b_h^2 N_{\bullet h\bullet}$. It follows that

$$SS_{\text{Total}} = \sum_{h=1}^{c} b_h^2 N_{\bullet h\bullet} - \frac{\left(\sum_{h=1}^{c} b_h N_{\bullet h\bullet}\right)^2}{n} = \frac{(n-1)S^2}{n}.$$

Since $V_i = T_i - \frac{n_i T}{n}$, we have that $\sum_{i=1}^{t} \frac{V_i^2}{n_i} = \sum_{i=1}^{t} \frac{T_i^2}{n_i} - CF = SS_{\text{Treat}}$. Now, on using $F = \frac{\frac{SS_{\text{Treat}}}{t-1}}{\frac{SS_{\text{Error}}}{n-t}}$

$$
\begin{aligned}
W_{\text{MS}} &= \frac{n \sum_{i=1}^{t} \frac{V_i^2}{n_i}}{S^2} = \frac{(n-1) SS_{\text{Treat}}}{SS_{\text{Total}}} \\
&= \frac{(n-1) SS_{\text{Treat}}}{SS_{\text{Treat}} + SS_{\text{Error}}} \\
&= \frac{(n-1) \frac{SS_{\text{Treat}}}{SS_{\text{Error}}}}{\left(\frac{SS_{\text{Treat}}}{SS_{\text{Error}}} + 1\right)} \\
&= \frac{(n-1)(t-1) F}{n - t + (t-1) F}.
\end{aligned}
$$

When rearranged this gives the same relationship as that between KW and F:

$$
F = \frac{(n-t) W_{\text{MS}}}{(t-1)\left(n - 1 - W_{\text{MS}}\right)}.
$$

3.7.3 Whiskey Data Revisited

Using the ranks of the whiskey grade, routine hand calculations show that the factor, error, and total sums of squares are $20\frac{5}{6}$, $16\frac{2}{3}$, and $37\frac{1}{2}$, respectively. These figures yield a value of $3\frac{1}{8}$ for the ANOVA F test statistic. Coincidently, this is the same value when using the raw data. With $n = 8$, $t = 3$, and $F = \frac{25}{8}$, $W_{\text{MS}} = \frac{(n-1)(t-1)F}{n-t+(t-1)F}$ gives $W_{\text{MS}} = \frac{35}{9}$, as previously. The $F_{2,5}$ p-value is 0.1317 compared with 0.1431 using χ_2^2.

3.7.4 Homosexual Marriage Data Revisited

Suppose now that these data are aggregated over education, so the design becomes a CRD. When the mid-ranks are used as responses standard software gives a value of the ANOVA F statistic of 10.0419. This is consistent with the adjusted Kruskal–Wallis statistic predicted by the relationship, 17.6639. The ANOVA F p-value using the $F_{2,130}$ distribution of 0.000 also agrees with the χ_2^2 p-value found previously.

3.7.5 Corn Data

In Rayner and Best (2001, pp. 45–46), the corn data given in Table 3.1 are discussed. The raw data are yields when corn is grown by four different methods. These data were divided into four approximately equal categories, and

Table 3.1 Corn data.

Method	Outcome				Total
	1	2	3	4	
1	0	3	4	2	$n_1 = 9$
2	1	6	3	0	$n_2 = 10$
3	0	0	1	6	$n_3 = 7$
4	8	0	0	0	$n_4 = 8$
Total	9	9	8	8	$n = 34$

the table records the counts for each method in each category. Strong evidence of both location and dispersion differences in the methods are found in Rayner and Best (2001).

The ANOVA F statistic with mid-ranks as responses takes the value 36.4851. The $F_{3,30}$ p-value is 0.000.

The data are heavily tied, with mid-ranks 5, 14, 22.5, and 30.5. The rank sums for treatments 1–4 are 193, 156.5, 205.5, and 40, respectively, giving a value of 24.2863 for the unadjusted Kruskal–Wallis statistic and 25.9009 for the adjusted Kruskal–Wallis statistic. The latter follows since for $t_g = 8$ and 9, $t_g^3 - t_g = 504$ and 720, respectively, giving

$$C_{\mathrm{CRD}} = 1 - \frac{\sum_{g=1}^{G} t_g^3 - t_g}{(n-1)\,n\,(n+1)} = 1 - \frac{2448}{33 \times 34 \times 35} = 0.937662.$$

Both the Kruskal–Wallis and adjusted Kruskal–Wallis tests have p-value 0.000. The null hypothesis of no treatment differences is rejected at all the usual levels of significance.

As a check of the calculations, the relationship between the ANOVA F statistic (36.4851) and KW_A is satisfied here.

```
KW(treatment = method, response = outcome)

##
##          Kruskal-Wallis Rank Sum Test
##
## data:  outcome and method
## Adjusting for ties:
## KW_A = 25.90095, df = 3, p-value = 9.662855e-06
##
## No adjustment for ties:
## KW = 24.28634, df = 3, p-value = 2.176699e-05
```

3.8 Simulation Study

It is well known that the usual χ^2 approximations to the sampling distributions of the Friedman and Durbin tests may be improved by using F distribution approximations to simple transformations of these statistics. See, for example, Best and Rayner (2014). However, Spurrier (2003) noted that previous studies had shown that the F approximation is quite liberal for small sample sizes and inferior to the χ^2 approximation. We believe these studies are largely concerned with data for which there are no ties. We now report our own simulation study in which we find that when there are ties, the F approximation is quite competitive.

As in Rayner and Livingston Jr. (2020), consider a CRD with t groups on sizes $n_1, \ldots, n_t$ with the total sample size being $n_1 + \cdots + n_t = n$. For an indicative study, we take $n_i = s$ for all i. Next, take a grid with $t = 3, 7, 12$, and 20 and $s = 4, 7$, and 10. This gives sample sizes between 12 and 200. At the upper end of this range, this should be sufficient for asymptotic approximations to be excellent.

A random sample of size $n = st$ was obtained as follows: generate n uniform $(0, 1)$ values and categorise these into c intervals of equal lengths. Rank these categorised values, assigning mid-ranks to ties, and randomly allocate these mid-ranks to the treatments. Calculate KW_A and reject the null hypothesis of no treatment effects for values greater than the critical point of the χ^2_{t-1} distribution, and similarly calculate the F statistic $F = \frac{(n-t)KW_A}{(t-1)(n-1-KW_A)}$ and reject the null hypothesis for values greater than the critical point of the $F_{t-1,n-t}$ distribution. The proportion of rejections in 1,000,000 samples with nominal significance level 0.05 was recorded.

We explored the effect of different values of c and found little change between small and large values. Note that smaller values of c mean heavy categorisation and hence many ties. We give results for only $c = 3, 10, 20, 50$, and 100, representing heavy categorisation becoming progressively lighter.

For heavy categorisation and small sample sizes, some values of zero occurred for C_{CRD}, giving indeterminate values of KW_A. These reflect the samples in which all observations were given the same rank. These samples were discarded and replaced. We feel that if such a sample was obtained in practice, it, too, would be discarded.

Our naive expectation was that the χ^2 approach would be superior in larger sample sizes, while F would do well for larger c, on the basis that if there are too few categories, the F statistic will have fewer achievable values, and hence, the data couldn't be considered to be consistent with normality. This would be balanced by the well-known robustness of the F test.

Table 3.2 Proportion of rejections using Kruskal–Wallis and F tests for a nominal significance level of 0.05 based on 1,000,000 simulations. The first entry in each cell uses the χ^2 sampling distribution, while the second uses the F sampling distribution.

$c = 3$ categories		Number of observations for each treatment (s)		
		4	7	10
Number of	3	0.0402/0.0546	0. 0445/0.0541	0. 0469/0.0526
treatments (t)	7	0.0292/0.0543	0. 0401/0.0526	0. 0437/0.0520
	12	0.0282/0.0525	0. 0384/0.0514	0. 0421/0.0510
	20	0. 0275/0.0516	0. 0377/0. 0505	0. 0413/0.0502

$c = 10$ categories		Number of observations for each treatment (s)		
		4	7	10
Number of	3	0.0383/0.0605	0.0442/0.0534	0.0461/0.0522
treatments (t)	7	0.0287/0.0536	0.0393/0.0520	0.0434/0.0518
	12	0.0276/0.0519	0.0380/0.0510	0.0420/0.0509
	20	0.0274/0.0515	0.0375/0.0507	0.0413/0.0503

$c = 20$ categories		Number of observations for each treatment (s)		
		4	7	10
Number of	3	0.0386/0.0610	0.0442/0.0536	0.0464/0.0524
treatments (t)	7	0.0289/0.0539	0.0395/0.0520	0.0430/0.0512
	12	0.0276/0.0522	0.0385/0.0513	0.0418/0.0507
	20	0.0272/0.0513	0.0374/0.0504	0.0418/0.0508

$c = 50$ categories		Number of observations for each treatment (s)		
		4	7	10
Number of	3	0.0389/0.0596	0.0439/0.0535	0.0456/0.0515
treatments (t)	7	0.0288/0.0538	0.0398/0.0522	0.0430/0.0514
	12	0.0272/0.0516	0.0385/0.0513	0.0419/0.0508
	20	0.0273/0.0512	0.0377/0.0507	0.0416/0.0503

$c = 100$ categories		Number of observations for each treatment (s)		
		4	7	10
Number of	3	0.0394/0.0589	0.0440/0.0536	0.0462/0.0522
treatments (t)	7	0.0288/0.0539	0.0395/0.0521	0.0429/0.0511
	12	0.0275/0.0520	0.0383/0.0511	0.0419/0.0506
	20	0.0272/0.0511	0.0381/0.0511	0.0414/0.0503

See Table 3.2 for our results. These suggest that in this study, the F approach is never inferior to the χ^2 approach and is always acceptable. The F proportions were always greater than 0.05, while those for χ^2 were always less. For a fixed number of observations for each treatment (columns in the table), the χ^2 proportions deteriorate while the F proportions improve. For a fixed number of treatments, both approaches almost always improve as the number of observations for each treatment (rows in the table) increases.

Our set-up reflects how data may be obtained in consumer studies, and we acknowledge that other set-ups are possible. In particular, our treatment deals with ties as a data analyst would in practice: if ties occur, mid-ranks are used and the test statistic is adjusted. We conclude that the F approach doesn't deserve its poor press, especially when applied to testing allowing for ties.

3.9 Wald Test Statistics in the CRD

3.9.1 The Wald Test Statistic of General Association for the CMH Design

As earlier, suppose that for $i = 1, \ldots, t$ and $h = 1, \ldots, c$, $N_{ihj} = 1$ if the jth observation on the ith treatment is assessed as being in the hth category, and zero otherwise. Observations are independent. Put

$$P\left(N_{ihj} = 1\right) = \theta_{ih} + \beta_h,$$

where the θ_{ih} reflect the inclination for treatment i to be categorised in the hth category over and above the consensus of the other treatments, and where β_h is the probability, averaging over treatments, that an observation is categorised into the hth category: $\beta_h = \frac{n_{\bullet h \bullet}}{n}$.

Put $\theta_i = \left(\theta_{i1}, \ldots, \theta_{ic}\right)^{\mathrm{T}}, i = 1, \ldots, t$ and $\theta^{\mathrm{T}} = \left(\theta_1^{\mathrm{T}}, \ldots, \theta_t^{\mathrm{T}}\right)$. The hypothesis of no treatment effects can be assessed by testing the null hypothesis H: $\theta = 0$ against the alternative hypothesis K: $\theta \neq 0$ using a Wald statistic of the form $\hat{\theta}^{\mathrm{T}} \widehat{\mathrm{Cov}}^{-1}\left(\hat{\theta}\right) \hat{\theta}$. The $\{\beta_h\}$ are nuisance parameters. If the covariance matrix is singular, a generalised inverse such as the Moore–Penrose may be used instead of the matrix inverse. In the usual Wald test statistic, the asymptotic covariance matrix of the maximum likelihood estimator $\hat{\theta}$ is used and in Cov $\left(\hat{\theta}\right)$ the unknown parameter θ is replaced by $\hat{\theta}$. Here Wald-type tests will be constructed; these use the exact covariance matrix with parameters evaluated under the null hypothesis. Note that an advantage of using Wald-type tests is that it is unnecessary to fully specify the θ parameter space. Since our focus is $\theta = 0$, we merely need that zero is an interior point of this space.

In the current CRD setting, the CMH GA test is a conditional test, assuming that after summing over replicates (j), all treatment totals and all response category totals are assumed to be known. Thus, the number of treatments of type i, n_i, and the number of responses in category h, $n_{\bullet h \bullet}$, are all known constants and not random variables. It follows that an extended hypergeometric model is appropriate. Thus, the probability function for $\{N_{ih \bullet}\}$ is

$$P\left(N_{ih\bullet} = n_{ih\bullet} \text{ for all } i \text{ and } h\right) = \frac{\prod_{i=1}^{t} n_i! \; \prod_{h=1}^{c} n_{\bullet h \bullet}!}{n! \; \prod_{i=1}^{t} \prod_{h=1}^{c} n_{ih\bullet}!}.$$

To develop appropriate tests of H against K, it is necessary to first consider constraints on the parameters and then develop estimators of them. The constraints on $\{\theta_{ih}\}$ reflect the dimensions of the parameter space of the alternative hypothesis. These in turn give, since the Wald statistic has asymptotic χ^2 distribution, the degrees of freedom associated with the test statistic.

By the definition of β_h, $\beta_h = \frac{n_{\bullet h \bullet}}{n}$ for $h = 1, \dots, c$. Since $\sum_{h=1}^{c} \beta_h = 1$ there are $c - 1$ algebraically independent β_h. Moreover,

$$n\beta_h = \sum_{i=1}^{t} \sum_{j=1}^{b} P\left(N_{ihj} = 1\right) = \sum_{i=1}^{t} n_i \left(\theta_{ih} + \beta_h\right) = n\beta_h + \sum_{i=1}^{t} n_i \theta_{ih}.$$

Thus, $\sum_{i=1}^{t} n_i \theta_{ih} = 0$ for all h. The jth replicate of treatment i must be categorised into one of the categories, so $\sum_{h=1}^{c} P\left(N_{ihj} = 1\right) = \sum_{h=1}^{c} \left(\theta_{ih} + \beta_h\right) = \theta_{i\bullet} + \beta_\bullet = 1$. Since $\beta_\bullet = 1$, $\theta_{i\bullet} = 0$ for all i. Thus, as well as there being $c - 1$ algebraically independent β_h, there are $(c - 1)(t - 1)$ algebraically independent θ_{ih}.

To estimate θ_{ih} equate $N_{ih\bullet}$ with its expectation, $n_i \left(\theta_{ih} + \beta_h\right)$, giving

$$n_i \left(\hat{\theta}_{ih} + \beta_h\right) = N_{ih\bullet}$$

and hence $\hat{\theta}_{ih} = \frac{N_{ih\bullet}}{n_i} - \beta_h$, $i = 1, \dots, t$. This approach identifies the $\hat{\theta}_{ih}$ as a method of moments estimators. In fact, they are also maximum likelihood estimators.

To evaluate the Wald-type test statistic, we need to develop the variances and covariances of the $\hat{\theta}_{ih}$. To this end note using standard distribution theory for the extended hypergeometric distribution, the covariance between $N_{ih\bullet}$ and $N_{i'h'\bullet}$ is

$$\text{Cov}\left(N_{ih\bullet}, N_{i'h'\bullet}\right) = \frac{n_i n_{\bullet h \bullet} \left(n\delta_{ii'} - n_{i'}\right) \left(\delta_{hh'} n - n_{\bullet h' \bullet}\right)}{n^2 (n - 1)}.$$

See, for example, Landis et al. (1978, 1979). Since $\beta_h = \frac{n_{\bullet h\bullet}}{n}$,

$$\text{Cov}\left(N_{ih\bullet}, N_{i'h'\bullet}\right) = \frac{n_i \beta_h \left(n\delta_{ii'} - n_{i'}\right)\left(\delta_{hh'} - \beta_{h'}\right)}{n-1}.$$

Henceforth, it will be assumed that the null hypothesis of no treatment effects is true, and that both $i \neq i'$ and $h \neq h'$. Using the early moments of the extended hypergeometric distribution,

$$\text{E}\left[N_{ih\bullet}\right] = n_i \beta_h,$$

$$\text{Var}\left(N_{ih\bullet}\right) = \frac{n_i \beta_h \left(1 - \beta_h\right)\left(n - n_i\right)}{n-1},$$

$$\text{Cov}\left(N_{ih\bullet}, N_{i'h\bullet}\right) = \frac{-n_i n_{i'} \beta_h \left(1 - \beta_h\right)}{n-1},$$

$$\text{Cov}\left(N_{ih\bullet}, N_{ih'\bullet}\right) = \frac{-n_i \beta_h \beta_{h'} \left(n - n_i\right)}{n-1}, \text{ and}$$

$$\text{Cov}\left(N_{ih\bullet}, N_{i'h'\bullet}\right) = \frac{n_i n_{i'} \beta_h \beta_{h'}}{n-1}.$$

Now, since $\hat{\theta}_{ih} = \frac{N_{ih\bullet}}{n} + \beta_h$ and $\beta_h = \frac{n_{\bullet h\bullet}}{n}$ are known, $\text{Cov}\left(\hat{\theta}_{ih}\right) = \text{Cov}\left(\frac{N_{ih\bullet}}{n}\right)$. Thus,

$$(n-1)\ \text{Var}\left(\hat{\theta}_{ih}\right) = \frac{n - n_i}{n_i}\beta_h \left(1 - \beta_h\right),$$

$$(n-1)\ \text{Cov}\left(\hat{\theta}_{ih}, \theta_{i'h}\right) = -\beta_h \left(1 - \beta_h\right),$$

$$(n-1)\ \text{Cov}\left(\hat{\theta}_{ih}, \theta_{ih'}\right) = -\frac{n - n_i}{n_i}\beta_h \beta_{h'}, \text{ and}$$

$$(n-1)\ \text{Cov}\left(\hat{\theta}_{ih}, \theta_{i'h'}\right) = \beta_h \beta_{h'}.$$

Write $\sigma_{hh} = \beta_h \left(1 - \beta_h\right)$, $\sigma_{hh'} = -\beta_h \beta_{h'}$ and $\Sigma = \{\sigma_{hk}\} = \text{diag}\left(\beta_h\right) - \{\beta_h \beta_{h'}\}$. Recall that $\theta_i = \left(\theta_1, \ldots, \theta_{ic}\right)^{\text{T}}$, $i = 1, \ldots, t$ and $\theta^{\text{T}} = \left(\theta_1^{\text{T}}, \ldots, \theta_t^{\text{T}}\right)$. The diagonal elements of $(n-1)\ \text{Cov}\left(\hat{\theta}_i\right)$ are $\frac{n - n_i}{n_i}\sigma_{hh}$ and the off-diagonal elements are $\frac{n - n_i}{n_i}\sigma_{hh'}$. So this matrix is $\frac{n - n_i}{n_i}\Sigma$. The diagonal elements of $(n-1)\ \text{Cov}\left(\hat{\theta}_i, \hat{\theta}_{i'}\right)$ are $-\sigma_{hh}$ and the off-diagonal elements are $-\sigma_{hh'}$. So this matrix is $-\Sigma$. If $\otimes$ is the Direct or Kronecker product, then

$$(n-1)\ \text{Cov}\left(\hat{\theta}\right) = \left[\text{diag}\left(\frac{n}{n_i}\right) - 1_t 1_t^{\text{T}}\right] \otimes \Sigma.$$

Since $\text{Cov}\left(\hat{\theta}\right)$ is not of full rank, we need its Moore–Penrose inverse, and hence, the Moore–Penrose inverse of both factors. First, write $D = \text{diag}\left(\sqrt{\frac{n}{n_i}}\right)$ and $u = D^{-1}1_t$. Note that $u^{\text{T}}u = 1$. With this notation, the first factor is $\text{diag}\left(\frac{n}{n_i}\right) - 1_t 1_t^{\text{T}} = D\left(I_t - D^{-1}1_t 1_t^{\text{T}}D^{-1}\right)D = D\left(I_t - uu^{\text{T}}\right)D$.

Now, $I_t - uu^T$ is idempotent, and the Moore–Penrose inverse of any idempotent matrix is itself. Hence,

$$\left[\text{diag} \left(\frac{n}{n_i} - 1_t 1_t^T \right) \right]^- = D^{-1} (I_t - uu^T)^- D^{-1}$$
$$= D^{-1} (I_t - uu^T) D^{-1}$$
$$= D^{-2} - D^{-1} uu^T D^{-1}$$
$$= \text{diag} \left(\frac{n_i}{n} \right) - \left\{ \frac{n_i n_j}{n^2} \right\}$$

as $D^{-1} u = D^{-2} 1_t = \frac{n_i}{n}$.

Next, we need the Moore–Penrose inverse of the second factor, $\Sigma = \text{diag}(\beta_h) - \{\beta_h \beta_{h'}\}$. In a similar manner, to the derivation in the previous paragraph the Moore–Penrose inverse of Σ is $\Sigma^- = \text{diag}(\beta_h^{-1}) - 1_c 1_c^T$. It follows that the Moore–Penrose inverse of $(n-1) \, \text{Cov}(\hat{\theta})$ is

$$\text{Cov}^- (\hat{\theta}) = (n-1) \left[\text{diag} \left(\frac{n_i}{n} \right) - \left\{ \frac{n_i n_j}{n^2} \right\} \right] \otimes \left[\text{diag}(\beta_h^{-1}) - 1_c 1_c^T \right].$$

To evaluate the quadratic form the test statistic, a little extra mathematics is useful. Note that the *vectorisation* of a matrix A is the vector formed by stacking the columns of A one upon the other: thus, if $A = (a_1, \dots, a_m)$ then $\text{vec}(A) = (a_1^T, \dots, a_m^T)^T$. A standard identity (see, for example, https://en.wikipedia.org/wiki/Vectorization_(mathematics)) is that

$$(B^T \otimes A) \, \text{vec}(X) = \text{vec}(AXB).$$

Now, write $\Theta = (\theta_1 | \dots | \theta_t)$ and $\hat{\Theta} = (\hat{\theta}_1 | \dots | \hat{\theta}_t)$. Then $\hat{\theta} = \text{vec}(\hat{\Theta})$. Putting $B = \text{diag} \left(\frac{n_i}{n} \right) - \left\{ \frac{n_i n_j}{n^2} \right\}$, $A = \Sigma^{-1}$, and $X = \hat{\Theta}$ in the previous identity gives

$$(B \otimes \Sigma^{-1}) \, \text{vec}(\hat{\Theta}) = \text{vec}(\Sigma^{-1} \hat{\Theta} B).$$

Now, $\hat{\Theta} B = \hat{\Theta} \text{diag} \left(\frac{n_i}{n} \right) - \hat{\Theta} \left\{ \frac{n_i n_j}{n^2} \right\}$. First, $\hat{\Theta} \text{diag} \left(\frac{n_i}{n} \right) = \left(\frac{n_1 \hat{\theta}_1}{n} | \dots | \frac{n_t \hat{\theta}_t}{n} \right)$ and second, $\hat{\Theta} n_i n_j = \left(\sum_{i=1}^t n_i \hat{\theta}_{ih} \right) n = 0$, since

$$\sum_{i=1}^t n_i \hat{\theta}_{ih} = \sum_{i=1}^t n_i \left(\frac{N_{ih\bullet}}{n_i} - \beta_h \right) = \sum_{i=1}^t n_i \left(\frac{N_{ih\bullet}}{n_i} - \frac{n_{\bullet h \bullet}}{n} \right) = 0.$$

It follows that

$$\text{vec}(\Sigma^{-1} \hat{\Theta} B) = \text{vec} \left(\Sigma^{-1} \left(\frac{n_1}{n} \hat{\theta}_1 | \dots | \frac{n_t}{n} \hat{\theta}_t \right) \right)$$
$$= \text{vec} \left(\frac{n_1}{n} \Sigma^{-1} \hat{\theta}_1 | \dots | \frac{n_t}{n} \Sigma^{-1} \hat{\theta}_t \right).$$

The test statistic is given by

$$
\begin{aligned}
W_{\text{GA}} = \hat{\theta}^{\text{T}} \text{Cov}^{-}\left(\hat{\theta}\right) \hat{\theta} &= (n-1) \hat{\theta}^{\text{T}} \left[\text{B} \otimes \Sigma^{-1}\right] \hat{\theta} \\
&= (n-1) \left(\hat{\theta}_1^{\text{T}}, \dots, \hat{\theta}_t^{\text{T}}\right)^{\text{T}} \left(\frac{n_1}{n} \Sigma^{-1} \hat{\theta}_1 | \dots | \frac{n_t}{n} \Sigma^{-1} \hat{\theta}_t\right) \\
&= \frac{n-1}{n} \sum_{i=1}^{t} n_i \hat{\theta}_i^{\text{T}} \Sigma^{-1} \hat{\theta}_i.
\end{aligned}
$$

Recall that $\hat{\theta}_{ih} = \frac{N_{ih\bullet}}{n_i} - \beta_h$ and $\beta_h = \frac{n_{\bullet h\bullet}}{n}$, so that $\Sigma^{-} \hat{\theta}_i = \left(\frac{N_{ih\bullet}}{n_i \beta_h} - 1\right)$ and

$$
\hat{\theta}_i^{\text{T}} \Sigma^{-1} \hat{\theta}_i = -1 + \frac{n}{n_i} \sum_{h=1}^{c} \frac{N_{ih\bullet}^2}{n_{\bullet h\bullet}}.
$$

The Pearson X_{P}^2 statistic of the form $-n + \sum \frac{\text{observed}^2}{\text{expected}}$ is given by

$$
X_{\text{P}}^2 = -n + n \sum_{i=1}^{t} \sum_{h=1}^{c} \frac{N_{ih\bullet}^2}{n_i n_{\bullet h\bullet}}
$$

and routine algebra shows that $\sum_{i=1}^{t} n_i \hat{\theta}_i^{\text{T}} \Sigma^{-1} \hat{\theta}_i = X_{\text{P}}^2$ so that $W_{\text{GA}} = \frac{n-1}{n} X_{\text{P}}^2$: the Wald test statistic is the CMH GA statistic given in Section 2.3.

By the central limit theorem (CLT) $\hat{\theta}$ is asymptotically normal. The quadratic form W_{GA} has asymptotic chi-squared distribution with degrees of freedom those of Cov $\left(\hat{\theta}\right)$. These are the product of those of the factor matrices in Cov $\left(\hat{\theta}\right)$, namely, $t-1$ in $I_t - \frac{1_t 1_t^{\text{T}}}{t}$ and $c-1$ in $\hat{\Sigma}$. Thus, W_{GA} has an asymptotic $\chi^2_{(c-1)(t-1)}$ distribution.

3.9.2 The Wald Test Statistic for the CMH MS Design

Suppose now that the score b_h is assigned to the hth response. We find the CMH MS test statistic for the particular case of a CRD. At the conclusion of this section, it will be shown that the corresponding test is a Wald test arising from a model related to that in the previous section.

Given the scores $\{b_h\}$, the score sum for treatment i is

$$
M_i = \sum_{h=1}^{c} \sum_{j=1}^{b} b_h N_{ihj} = \sum_{h=1}^{c} b_h N_{ih\bullet}.
$$

Using the extended hypergeometric distribution M_i has expectation

$$
\text{E}\left[M_i\right] = n_i \sum_{h=1}^{c} b_h \frac{n_{\bullet h\bullet}}{n}
$$

under the null hypothesis of no mean effects. Note that

$$\sum_{i=1}^{t} \left[M_i - \text{E}\left[M_i \right] \right] = \sum_{h=1}^{c} b_h n_{\bullet h \bullet} - n \sum_{h=1}^{c} b_h \frac{n_{\bullet h \bullet}}{n} = 0.$$

If $M = \{ M_i \}$, then inference about the population mean counts can be based on M through the quadratic form

$$S_{\text{MS}} = (M - \text{E}\left[M \right])^{\text{T}} \text{Cov}^- (M) (M - \text{E}\left[M \right]),$$

in which all unknown parameters are estimated by maximum likelihood under the null hypothesis. The asymptotic distribution of S_{MS} is χ_{t-1}^2. Since $\text{Cov}\,(M)$ is not of full rank, the Moore–Penrose generalised inverse is used in S_{MS}. This is the CMH MS test statistic. In the particular case of a CRD, it is of interest to specify $\text{Cov}\,(M)$ and thereby find the expression for the test statistic W_{MS} that is, not surprisingly, simpler than for S_{MS}.

To specify the CMH MS test statistic requires $\text{Cov}\,(M)$. The derivation of $\text{Cov}\left(M_i \right)$ requires the covariance between $N_{ih\bullet}$ and $N_{i'h'\bullet}$ given in the previous section. If we write $(n-1)S^2 = n \sum_{h=1}^{c} b_h^2 n_{\bullet h \bullet} - \left(\sum_{h=1}^{c} b_h n_{\bullet h \bullet} \right)^2$, then it follows that

$$\text{Var}\left(\sum_{h=1}^{c} b_h N_{ih\bullet} \right) = \frac{n_i \left(n - n_i \right) S^2}{n^2 (n-1)}$$

$$\text{Cov}\left(\sum_{h=1}^{c} b_h N_{ih\bullet}, b_h' N_{i'h'\bullet} \right) = -\frac{n_i n_{i'} S^2}{n^2 (n-1)}.$$

Now, define $p_i = \frac{n_i}{n}$. Then

$$\text{Cov}\,(M) = S^2 \left(\text{diag}\left(p_i \right) - \{ p_i p_{i'} \} \right).$$

Next, put $D = \text{diag}\left(\sqrt{p_i} \right)$ and $u = \left\{ \sqrt{p_i} \right\}$. Thus, $\text{Cov}\,(M) = S^2 D \left(I_t - uu^{\text{T}} \right) D$.

Since $u^{\text{T}} u = 1$, uu^{T} and $I_t - uu^{\text{T}}$ are idempotent, and as any idempotent matrix is its own Moore–Penrose inverse,

$$\text{Cov}^- (M) = \frac{D^{-1} \left(I_t - uu^{\text{T}} \right) D^{-1}}{S^2}.$$

If $V_i = M_i - \text{E}\left[M_i \right] = \sum_{h=1}^{c} b_h N_{ih\bullet} - n_i \sum_{h=1}^{c} b_h \frac{n_{\bullet h \bullet}}{n}$ and $V = \{ V_i \}$, then clearly $1_t^{\text{T}} V = 0$. Now,

$$W_{\text{MS}} = V^{\text{T}} \text{Cov}^- (M) V = \frac{V^{\text{T}} D^{-1} \left(I_t - uu^{\text{T}} \right) D^{-1} V}{S^2}$$

and since $u^{\text{T}} D^{-1} V = 1_t^{\text{T}} V = 0$,

$$W_{\text{MS}} = \frac{V^{\text{T}} D^{-2} V}{S^2} = \frac{n}{S^2} \sum_{i=1}^{t} \frac{V_i^2}{n_i}.$$

This is just a weighted sum of squares of the centred score sums. This is the CMH MS test statistic for the CRD.

In Section 3.9.1, we had $\Theta = \{\theta_{ih}\}$ has maximum likelihood estimator $\hat{\Theta} = \{\hat{\theta}_{ih}\}$ in which $\hat{\theta}_{ih} = \frac{N_{ih\bullet}}{n_i} - \beta_h$ and $\beta_h = \frac{n_{\bullet h\bullet}}{n}$. Now write $s_b = \{b_h\}$ for the block scores. Then $\hat{\Theta}s_b$ has elements

$$\sum_{h=1}^{c} b_h \left(\frac{N_{ih\bullet}}{n_i} - \frac{n_{\bullet h\bullet}}{n} \right) = \frac{(M_i - \mathrm{E}\,[M_i])}{n_i}.$$

If $\Lambda = \mathrm{diag}\,(n_i)$ and $M = \{M_i\}$, then $\hat{\Theta}s_b = \Lambda^{-1}(M - \mathrm{E}\,[M])$. Since W_{MS} is a quadratic form based on $(M - \mathrm{E}\,[M])$, it is also a quadratic form based on $\Lambda\hat{\Theta}s_b$. It is thus the Wald-type test statistic for testing $\Lambda\Theta s_b = 0$ against $\Lambda\Theta s_b \neq 0$. It thus enjoys the properties of Wald-type tests: the test statistic has asymptotic distribution χ^2, and the test is weakly optimal.

3.9.3 The Wald Test Statistic for the CMH C Design

In Section 2.5.2, we found that the CMH C test statistic was given by

$$W_C = \frac{C^2}{\mathrm{Var}\,(C)} = \frac{(n-1)\,S_{XY}^2}{S_{XX}S_{YY}} = (n-1)\,r_P^2,$$

in which r_P is the Pearson correlation coefficient.

Previously in this section, we had $s_b = \{b_h\}$ and that $\Theta = \{\theta_{ih}\}$ has maximum likelihood estimator $\hat{\Theta} = \{\hat{\theta}_{ih}\}$ in which $\hat{\theta}_{ih} = \frac{N_{ih\bullet}}{n_i} - \beta_h$ and $\beta_h = \frac{n_{\bullet h\bullet}}{n}$. Now, write $s_t = \{a_i\}$ for the treatment scores. Then $s_t^\mathrm{T}\Lambda\hat{\Theta}s_b$ is given by

$$s_t^\mathrm{T}\Lambda\hat{\Theta}s_b = \sum_{i=1}^{t}\sum_{h=1}^{c} a_i b_h \left(N_{ih\bullet} - \frac{n_i n_{\bullet h\bullet}}{n} \right) = C.$$

Since $W_C = \frac{C^2}{\mathrm{Var}(C)}$, if $\phi = s_t^\mathrm{T}\Lambda\hat{\Theta}s_b$ it is also $\frac{\hat{\phi}^2}{\hat{\mathrm{Var}}(\hat{\phi})}$. It is thus the Wald-type test statistic for testing $\phi = 0$ against $\phi \neq 0$. It thus enjoys the previously mentioned properties of Wald-type tests.

Bibliography

Best, D. J. and Rayner, J. C. W. (2014). Conover's F test as an alternative to Durbin's test. *Journal of Modern Applied Statistical Methods*, 13(2):4.

Davis, C. S. (2002). *Statistical methods for the analysis of repeated measurements*. Technical report, Springer.

Gibbons, J. D. and Chakraborti, S. (2021). *Nonparametric Statistical Inference*, 6th edition. Boca Raton, FL: CRC Press.

Landis, J. R., Cooper, M. M., Kennedy, T., and Koch, G. G. (1979). A computer program for testing average partial association in three-way contingency tables (PARCAT). *Computer Programs in Biomedicine*, 9(3):223–246.

Landis, J. R., Heyman, E. R., and Koch, G. G. (1978). Average partial association in three-way contingency tables: a review and discussion of alternative tests. *International Statistical Review/Revue Internationale de Statistique*, 46(3):237–254.

Rayner, J. C. W. and Best, D. J. (2001). *A Contingency Table Approach to Nonparametric Testing*. Boca Raton, FL: Chapman & Hall/CRC.

Rayner, J. C. W. and Livingston Jr., G. C. (2020). The Kruskal-Wallis tests are Cochran-Mantel-Haenszel mean score tests. *METRON*, 78(3):353–360.

Spurrier, J. D. (2003). On the null distribution of the Kruskal-Wallis statistic. *Journal of Nonparametric Statistics*, 15(6):685–691.

Thas, O., Best, D. J., and Rayner, J. C. W. (2012). Using orthogonal trend contrasts for testing ranked data with ordered alternatives. *Statistica Neerlandica*, 66(4):452–471.

4

The Randomised Block Design

4.1 Introduction

From some perspectives, the Cochran–Mantel–Haenszel (CMH) design could have been constructed to be applicable to randomised block data with categorical responses. In both the CMH design and the randomised block design (RBD), categorical responses are recorded when at least two treatments are applied to at least two blocks or strata. The RBD constrains the treatments so that precisely one of each treatment is applied on each block. As discussed in Section 1.1, many data sets are of three types: normal, ranked, and categorical. Here we make use of the fact that the last two overlap: ordered categorical responses can be scored by their mid-ranks.

The sequence of the material in this chapter will be similar to that for Chapter 3. First, the parametric model will be given; many data analysts would not use nonparametric methods unless the parametric assumptions are not obviously met. Hence, it is important to know what those assumptions are. Ranking methods are frequently used when the parametric assumptions are dubious and lead to the Friedman tests, for both when the data are distinct, and when there are ties. Next, the CMH tests will be developed for the particular scenario of the RBD, for then the test statistics take simpler forms than the most general.

In this chapter, we do not attempt to directly relate the Friedman and ANOVA F tests, although such a relationship exists. In Chapter 5, we will relate the Durbin and ANOVA F tests, and the relationship between the Friedman and ANOVA F tests is a particular case of that relationship.

In Section 4.4, we show that the Friedman tests are CMH tests, and since the CMH mean scores (MS) test statistic is simply related to the ANOVA F test statistic, so are the Friedman test statistics. Finally, we show for the RBD that the CMH general association (GA) test is a Wald-type test, whereas for the CMH MS and C tests, we can only establish that they are Wald tests

An Introduction to Cochran–Mantel–Haenszel Testing and Nonparametric ANOVA,
First Edition. J.C.W. Rayner and G. C. Livingston Jr.
© 2023 John Wiley & Sons Ltd. Published 2023 by John Wiley & Sons Ltd.

when both there are no ties and scores are not block dependent. Wald tests have weak optimality properties, and hence so do the Friedman tests that are Wald tests.

4.2 The Design and Parametric Model

In the RBD, we observe random variables Y_{ij}, $i = 1, \ldots, t$ and $j = 1, \ldots, b$. There are t treatments, each observed once on each of b blocks. Blocks are assumed to be inherently different, and observing the treatments in this way removes the block variability from contaminating the error, thus giving a sharper test for treatment effects. Blocks may be particular animals, known to be different to each other. The treatments may be different drugs. The sense of the null hypothesis is that the treatments have no location effect.

The fixed effects parametric model is

$$Y_{ij} = \mu + A_i + B_j + E_{ij},$$

in which $i = 1, \ldots, t$ and $j = 1, \ldots, b$ and where

- $\mu = \mathrm{E}[Y]$ is the overall average,
- A_i is the main effect of factor A, or the average (over j) effect of factor A at level i,
- B_j is the block effect at its jth level,
- E_{ij} is the random effect of Y_{ij}.

Moreover, the E_{ij} are normally distributed and mutually independent, each with mean 0 and variance σ^2, and

- the A_i and B_j are all constants satisfying $\sum_{i=1}^{t} A_i = \sum_{j=1}^{b} B_j = 0$.

Put $CF = \sum_{i=1}^{t} \sum_{j=1}^{b} \frac{Y_{ij}^2}{bt}$. Then the factor (treatment), block, and total sums of squares are, respectively,

$$SS_{\text{Treat}} = \sum_{i=1}^{t} \frac{Y_{i\bullet}^2}{b} - CF,$$

$$SS_{\text{Block}} = \sum_{j=1}^{b} \frac{Y_{\bullet j}^2}{t} - CF,$$

$$SS_{\text{Total}} = \sum_{i=1}^{t} \sum_{j=1}^{b} Y_{ij}^2 - CF,$$

with $t-1$, $b-1$ and $(bt-1)$ degrees of freedom, respectively. The error sum of squares and error degrees of freedom (edf) are found by difference: $SS_{\text{Error}} = SS_{\text{Total}} - SS_{\text{Treat}} - SS_{\text{Block}}$ and edf $= (bt-1) - (t-1) - (b-1) = (b-1)(t-1)$. The mean squares are the sums of squares divided by their degrees of freedom:

$$MS_{\text{Treat}} = \frac{SS_{\text{Treat}}}{t-1}, \quad MS_{\text{Block}} = \frac{SS_{\text{Block}}}{b-1}, \quad \text{and } MS_{\text{Error}} = \frac{SS_{\text{Error}}}{(b-1)(t-1)}.$$

The ANOVA F test statistic for testing for treatment effects is $\frac{MS_{\text{Treat}}}{MS_{\text{Error}}}$ and has sampling distribution $F_{t-1,\,(b-1)(t-1)}$. Block effects may be assessed similarly, although they are rarely of interest. The null hypothesis under this model is that the treatment means are all equal: $A_1 = \cdots = A_t = 0$.

4.3 The Friedman Tests

If the assumptions for the ANOVA F test for the RBD (normality, constant variance) are not satisfied, an alternative test that doesn't make these assumptions is needed. Conventional wisdom is that the Friedman test assesses whether or not the treatment distributions are consistent, and that it is known to be sensitive to differences in the means of the treatment ranks. We revisit this issue subsequently.

Suppose we have distinct (untied) observations, y_{ij}, with a particular y_{ij} being the response for ith of t treatments on the jth of b blocks. The observations are ranked within each block and R_i, the sum of the ranks for treatment i over all blocks is calculated for $i = 1, \ldots, t$. The Friedman test statistic is

$$Fdm = \frac{12}{bt(t+1)} \sum_{i=1}^{t} \left[R_i - \frac{b(t+1)}{2} \right]^2$$

$$= -3b(t+1) + \frac{12}{bt(t+1)} \sum_{i=1}^{t} R_i^2.$$

If ties occur, the test statistic needs adjusting. A general adjustment is

$$Fdm_A = \frac{(t-1)\left[\sum_{i=1}^{t} R_i^2 - \frac{b^2 t(t+1)^2}{4} \right]}{\sum_{i=1}^{t} \sum_{j=1}^{b} r_{ij}^2 - \frac{bt(t+1)^2}{4}},$$

in which r_{ij} is the rank of treatment i on block j. If there are no ties, then Fdm_A reduces to Fdm. For then on block j, the ranks are $1, 2, \ldots, t$.

To show Fdm_A reduces to Fdm, we need the identities (3.1) from Chapter 3, reproduced here again

$$1 + \cdots + n = \frac{n(n+1)}{2}$$

$$1^2 + \cdots + n^2 = \frac{n(n+1)(2n+1)}{6}$$

$$\frac{n(n+1)(2n+1)}{6} - n\left(\frac{n+1}{2}\right)^2 = \frac{(n-1)(n+1)}{12}.$$

Now, $\sum_{i=1}^{t} \sum_{j=1}^{b} r_{ij}^2 = \frac{bt(t+1)(2t+1)}{6}$ and the denominator in Fdm_A is $\frac{b(t-1)t(t+1)}{12}$. With this simplification Fdm_A reduces to Fdm. As in Chapter 3, the basis for the adjustment is to replace the variance of the untied ranks on each block, $\frac{(t-1)(t+1)}{12}$, by the variance of the tied ranks, $\sum_{i=1}^{t} \frac{r_{ij}^2}{t} - \left(\frac{t+1}{2}\right)^2$. This assumes the sum of the ranks on each block is $\frac{t(t+1)}{2}$, as for the untied data.

An alternative expression is available for the adjusted test statistic for the particular scenario when there are ties and mid-ranks are used. This expression uses the number of tied data in a group of mid-ranks and is easier to calculate by hand, especially when there are few ties. With this adjustment, the Friedman statistic will be denoted by Fdm_M. However, Fdm_A is more general; it applies whenever there are ties that preserve the rank sums – mid-ranks are *not* assumed.

As when discussing the completely randomised design (CRD) in Section 3.3, when passing from untied to tied data using mid-ranks for n observations, the sum of squares is reduced by the aggregation of the corrections. Suppose that on the jth block the gth of G groups is of size t_{gj}. The aggregation of all such corrections is $\sum_{g=1}^{G} \sum_{j=1}^{b} \left(t_{gj}^3 - t_{gj}\right)$ for all groups of tied observations. The denominator in the Friedman test statistic becomes

$$b\left(1^2 + \cdots + t^2\right) - \sum_{g=1}^{G} \sum_{j=1}^{b} \frac{t_{gj}^3 - t_{gj}}{12} - b\left[\frac{t(t+1)}{2}\right]^2$$

$$= \frac{b(t-1)t(t+1)}{12} - \sum_{g=1}^{G} \frac{t_g^3 - t_g}{12} = C_{RBD}\frac{b(t-1)t(t+1)}{12},$$

in which

$$C_{RBD} = 1 - \frac{\sum_{g=1}^{G} \sum_{j=1}^{b} \left(t_{gj}^3 - t_{gj}\right)}{b(t-1)t(t+1)}.$$

Thus $Fdm_M = \frac{Fdm}{C_{RBD}}$. The asymptotic distributions of Fdm_A, Fdm_M, and Fdm_A are all χ_{t-1}^2.

4.3.1 Jams Data

Three plum jams, A, B, and C were given just about right (JAR) sweetness codes by eight judges. The codes ranged from 1, denoting 'not sweet enough' to 5, 'too sweet'. The CMH tests for these data were applied in Section 2.2 and were previously analysed in Rayner and Best (2017, 2018). Table 2.2 gives the ranked data. The raw (unranked) data are in parentheses. Judges are blocks.

Using mid-ranks the rank sums are 12.5, 19.5, and 16, leading to Friedman test statistic of $Fdm = \frac{49}{16} = 3.0625$. On six blocks there are two ties, so that on these blocks $\left(t_{gj}^3 - t_{gj}\right) = 6$, leading to an adjustment factor of $C_{RBD} = \frac{13}{16}$ and an adjusted Friedman statistic $Fdm_M = \frac{Fdm}{C_{RBD}} = \frac{49}{13} = 3.7692$. Alternatively, the sum of the squares of the mid-ranks is 109 and substituting leads to $Fdm_A = 3.7692$.

The adjusted Friedman test has a χ_2^2 p-value of 0.152. This compares with p-value 0.216 for the unadjusted Friedman test using the χ_2^2 distribution.

Calculating directly, the ANOVA F statistic F calculated from the ranked data is 2.1572 with $F_{2,27}$ p-value 0.153. This will be briefly revisited in Section 4.6.

```
friedman(y = sweetness,groups = type,blocks = judge)

##
##          Friedman Rank Sum Test
##
## Test statistic is adjusted for ties
## data:   sweetness type judge
## Fdm_A = 3.769231, df = 2, p-value = 0.1518875
##
## For the group levels A B C,
## the rank sums are:   12.5 19.5 16


Anova(lm(sweetness_ranks~type+judge), type = 3)

## Anova Table (Type III tests)
##
## Response: sweetness_ranks
##              Sum Sq Df F value  Pr(>F)
## (Intercept)  5.8594  1  8.2547 0.01228 *
## type         3.0625  2  2.1572 0.15252
## judge        0.0000  7  0.0000 1.00000
## Residuals    9.9375 14
## ---
## Signif. codes: 0 '***' 0.001 '**' 0.01 '*' 0.05 '.' 0.1 ' ' 1
```

4.4   The CMH Test Statistics in the RBD

Compared with the CRD, slightly different notation is needed to define the CMH test statistics for the RBD. Here for $i = 1, \ldots, t$, $h = 1, \ldots, c$, and $j = 1, \ldots, b$ suppose that $N_{ihj} = 1$ if an observation is assessed to be of the ith treatment in the hth category on the jth block, and zero otherwise. For this design, every treatment is observed precisely once on each block: $n_{i \bullet j} = 1$ for all i and j. Consequently, $n_{\bullet \bullet j} = t$, $n_{i \bullet \bullet} = b$, $n_{\bullet \bullet \bullet} = bt$, and $p_{i \bullet j} = \frac{n_{i \bullet \bullet}}{n_{\bullet \bullet \bullet}} = \frac{1}{t}$. These constraints show that inference here is conditional.

4.4.1   The CMH OPA Test for the RBD

In Section 2.3, the CMH overall partial association (OPA) test statistic was described to be the sum over strata of weighted Pearson X^2 statistics, $X_{\mathrm{P}j}^2$. Here all weights are $\frac{t-1}{t}$. There is no further simplification available for this test.

An alternative test for OPA, not based on conditional inference, could be based on $\sum_{j=1}^{b} X_{\mathrm{P}j}^2$. The null and alternative hypotheses are the same as for the CMH OPA test, as is the asymptotic distribution of the test statistic. This is discussed again briefly in Section 6.1.

4.4.2   The CMH GA Test Statistic for the RBD

The CMH GA test statistic is a quadratic form with vector having elements $\{N_{ih \bullet}\}$: the counts aggregated over strata. An alternative test, testing the same null and alternative hypotheses and with the same asymptotic distribution, is the Pearson statistic for the table of counts $\{N_{ih \bullet}\}$, namely,

$$X_{\mathrm{P} \bullet}^2 = \sum_{i=1}^{t} \sum_{j=1}^{b} \frac{\left(N_{ih \bullet} - \frac{n_{i \bullet \bullet} n_{\bullet h \bullet}}{n_{\bullet \bullet \bullet}}\right)^2}{\frac{n_{i \bullet \bullet} n_{\bullet h \bullet}}{n_{\bullet \bullet \bullet}}}.$$

In Section 4.8, we derive the Wald test statistic W_{GA} in what might be called the GA CMH design and show that $S_{\mathrm{GA}} = W_{\mathrm{GA}}$. This Wald statistic is given by

$$W_{\mathrm{GA}} = \frac{t-1}{t} \sum_{i=1}^{t} d_i^{\mathrm{T}} V^{-1} d_i,$$

in which d_i has hth element $N_{ih \bullet} - \frac{N_{\bullet h \bullet}}{t}$ for $i = 1, \ldots, t$ and $h = 1, \ldots, (c-1)$. To avoid V being singular the cth product category, which is redundant, has been omitted. To simplify hand calculation of V, form the

- $t \times (c-1)$ treatments by responses counts matrix, $R = \{N_{ih \bullet}\}$
- $b \times (c-1)$ blocks by responses counts matrix, $S = \{N_{\bullet hj}\}$
- $(c-1) \times (c-1)$ diagonal matrix $T = \mathrm{diag}\left(N_{\bullet h \bullet}\right)$.

Then $V = \frac{T}{t} - \frac{S^T S}{t^2}$. Of course, the $N_{\bullet h \bullet}$ can be obtained by summing the $N_{ih\bullet}$ over treatments or the $N_{\bullet hj}$ over blocks. Often, the $N_{ih\bullet}$ and $N_{\bullet hj}$ can be written down from the data, and then S_{GA} calculated from the expression earlier. With this approach, there is no need to enter the full data set into standard software.

4.4.3 The CMH MS Test Statistic for the RBD

Previously, we used the notation S_{MS} for the CMH tests in general. In Chapter 3, CMH tests for the CRD were denoted by W_{MS} etc. As there will be no ambiguity, we will again use W_{MS} for CMH MS test for the RBD.

To give the CMH MS test, recall some general results from Section 2.3:

$$S_{MS} = V^T \text{Cov}^- (M) V \text{ in which } V = \left\{ M_{i\bullet} - \sum_{j=1}^{b} n_{i\bullet j} \sum_{h=1}^{c} \frac{b_{hj} n_{\bullet hj}}{n_{\bullet \bullet j}} \right\},$$

and if

$$p_{i\bullet j} = \frac{n_{i\bullet j}}{n_{\bullet \bullet j}}, \quad D_j = \text{diag}\left(\sqrt{p_{i\bullet j}}\right), \quad u_j = \left\{\sqrt{p_{i\bullet j}}\right\}, \text{ and}$$

$$S_j^2 = \frac{n_{\bullet \bullet j} \sum_{h=1}^{c} b_{hj}^2 n_{\bullet hj} - \left(\sum_{h=1}^{c} b_{hj} n_{\bullet hj}\right)^2}{n_{\bullet \bullet j} - 1},$$

then $\text{Cov}(M) = \sum_{j=1}^{b} \text{Cov}(M_j) = \sum_{j=1}^{b} S_j^2 D_j \left(I_t - u_j u_j^T\right) D_j$.

These quantities simplify for the RBD, since, as before,

$$n_{i\bullet j} = 1, \ n_{\bullet \bullet j} = t, \ n_{i\bullet\bullet}) = b, n_{\bullet\bullet\bullet} = bt.$$

Then

$$p_{i\bullet j} = \frac{n_{i\bullet\bullet}}{n_{\bullet\bullet\bullet}} = \frac{1}{t}, \quad D_j = \frac{I_t}{\sqrt{t}}, \quad u_j = \frac{1_t}{\sqrt{t}}, \quad I_t - u_j u_j^T = I_t - \frac{1_t 1_t^T}{t},$$

and hence

$$D_j \left(I_t - u_j u_j^T\right) D_j = \frac{I_t - \frac{1_t 1_t^T}{t}}{t}.$$

This gives $\text{Cov}(M) = \frac{\left(\sum_{j=1}^{b} S_j^2\right)\left(I_t - \frac{1_t 1_t^T}{t}\right)}{t}$. The Moore–Penrose inverse of an idempotent matrix is itself, so $\text{Cov}^-(M) = \frac{t\left(I_t - \frac{1_t 1_t^T}{t}\right)}{\sum_{j=1}^{b} S_j^2}$. It now follows that

$$V_i = M_{i\bullet} - \text{E}\left[M_{i\bullet}\right] = \sum_{h=1}^{c} b_{hj} N_{ih\bullet} - \sum_{h=1}^{c} \frac{b_{hj} n_{\bullet h\bullet}}{t},$$

and since $1_t^T V = 0$, W_{MS}, the CMH MS test statistic for the RBD, is given by

$$W_{MS} = \frac{t \sum_{i=1}^{t} V_i^2}{\sum_{j=1}^{b} S_j^2},$$

in which

$$(t-1)\,S_j^2 = t\sum_{h=1}^{c} b_{hj}^2 n_{\bullet hj} - \left(\sum_{h=1}^{c} b_{hj} n_{\bullet hj}\right)^2.$$

In Section 4.6, it will be shown that for randomised blocks

$$W_{\mathrm{MS}} = \frac{b(t-1)F}{b-1+F},$$

in which F is the F test statistic from the randomised block ANOVA using the data $\{b_{hj}N_{ihj}\}$. The asymptotic distribution of W_{MS} is χ_{t-1}^2, while the exact distribution of F is $F_{t-1,\,(b-1)(t-1)}$, provided the ANOVA assumptions are met. In Section 4.7, we give a simulation study that favours use of the F distribution. However, for highly categorised data, normality of the ANOVA residuals must be suspect, whatever a goodness of fit test for normality indicates. On the other hand, the χ_{t-1}^2 of W_{MS} distribution is asymptotic; how large does b need to be for this to be a good approximation?

4.4.3.1 Food Products Data

Table 4.1 gives cross-classified counts based on categorical responses made by 15 subjects for two different prices of the same food product. Table 4.2 gives the same data in a $t \times c$ layout. In this table, we use the code 1 for 'buy', code 2 for 'undecided', and code 3 for 'not buy'. The codes are not scores. Here $t = 2$, $c = 3$, and $b = 15$. The two products are price 1 and price 2. We now calculate the various CMH test statistics.

Table 4.1 Cross-classified counts of categorical responses made by 15 subjects for two different prices of the same food product.

	Price 2		
Price 1	**Buy**	**Undecided**	**Not buy**
Buy	6	2	1
Undecided	0	4	0
Not buy	0	0	2

Table 4.2 Long form of the same data in Table 4.1.

Subject	1	2	3	4	5	6	7	8	9	10	11	12	13	14	15
Price 1	1	1	1	1	1	1	1	1	1	2	2	2	2	3	3
Price 2	1	1	1	1	1	1	2	2	3	2	2	2	2	3	3

Table 4.3 Observed frequencies with expected frequencies in parentheses for blocks 7 and 8 of the food products data.

Treatment/outcome	1	2	3	Total
1	1 (0.5)	0 (0.5)	0 (0)	1
2	0 (0.5)	1 (0.5)	0 (0)	1
Total	1	1	0	2

For blocks 1–6 and 10–15, the Pearson X^2s are all zero. For blocks 7 and 8, the observed and expected tables are as in Table 4.3. Blocks 7–9 have Pearson X^2 all 0.5. Thus, $\sum_{j=1}^{15} X_{Pj}^2 = 1.5$. Since $n_{\bullet\bullet j} = 2$ for all j, $W_{OPA} = \sum_{j=1}^{15} \frac{(n_{\bullet\bullet j}-1)X_{Pj}^2}{n_{\bullet\bullet j}} = 0.75$. Both $\sum_{j=1}^{15} X_{Pj}^2$ and W_{OPA} are clearly not significant at the usual levels of significance.

When aggregated over blocks the observed and expected counts are as in Table 4.4. A simple calculation gives $X_{P\bullet}^2 = 1.1$, which is again not significant at the usual levels of significance (χ^2 p-value 0.577).

To calculate W_{GA} for these data, it is first necessary to calculate R, S, T, and d_1. The first row of R reflects that nine times price 1 is given outcome 1, which is to buy, then four times it is given outcome 2, undecided, and twice it is given outcome 3, don't buy. The second row reflects the outcomes for the second price. The matrix S has 15 rows, corresponding to the subjects or blocks. The first six subjects all give two outcomes buy and no outcomes undecided. The next two give one outcome buy and one undecided, and so on. To form the d_i first note that $N_{\bullet 1\bullet} = 15$, $N_{\bullet 2\bullet} = 10$, and $N_{\bullet 3\bullet} = 5$ so that d_1 has first element $9 - \frac{15}{2}$ and second element $4 - \frac{10}{2}$. Moreover, for $t = 2$, $d_1 + d_2$ always equals zero, so that $d_1 = -d_2$.

Table 4.4 Observed frequencies with expected frequencies in parentheses for the aggregated food products data.

Treatment/outcome	1	2	3	Total
1	9 (7.5)	4 (5)	2 (2.5)	15
2	6 (7.5)	6 (5)	3 (2.5)	15
Total	15	10	5	30

We find

$$R = \begin{pmatrix} 9 & 4 & 2 \\ 6 & 6 & 3 \end{pmatrix}, \quad S^T = \begin{pmatrix} 2 & 2 & 2 & 2 & 2 & 2 & 1 & 1 & 1 & 0 & 0 & 0 & 0 & 0 & 0 \\ 0 & 0 & 0 & 0 & 0 & 0 & 1 & 1 & 0 & 2 & 2 & 2 & 2 & 0 & 0 \end{pmatrix},$$

$$d_1^T = \begin{pmatrix} \frac{3}{2}, -1 \end{pmatrix}, \quad T = \begin{pmatrix} 15 & 0 \\ 0 & 10 \end{pmatrix}, \quad \text{and } V = \begin{pmatrix} \frac{3}{4} & -\frac{1}{2} \\ -\frac{1}{2} & \frac{1}{2} \end{pmatrix}$$

giving $W_{GA} = d_1^T V^{-1} d_1 = 3.00$ with χ_2^2 p-value 0.22.

If the codes are now interpreted as scores and the responses are analysed parametrically as a RBD, we find treatments are not significant (p-value 0.104), while the Shapiro–Wilk test for normality of the residuals has p-value 0.001. Since the treatment F statistic takes the value 3.027, using $W_{MS} = \frac{b(t-1)F}{b-1+F}$ gives $W_{MS} = 2.6667$ with χ_1^2 p-value 0.102. In spite of the responses being highly categorised and the residuals highly nonnormal, the χ^2 and F test p-values are remarkably similar. It seems there is no evidence of a mean treatment effect at significance levels up to 10%.

```
b_hj = matrix(1:3, ncol = 15, nrow = 3)
CMH(treatment = price, response = decision,
      strata = subject, b_hj = b_hj, test_OPA = FALSE,
      test_C = FALSE)
```

```
##
##          Cochran Mantel Haenszel Tests
##
##                           S df p-value
## General Association 3.000   2  0.2231
## Mean Score          2.667   1  0.1025
```

4.4.4 The CMH C Test Statistic for the RBD

The CMH C test statistic was expressed in an informative and useful form in Section 2.4, with a further simplification for the RBD in Section 2.5.4. As with the CMH OPA test, in the case of the RBD, there is no further simplification available. However, we recall that material through the following example.

4.4.4.1 Human Resource Data

Applications for a position are vetted and ranked by human resource (HR) professionals. The top five are interviewed by a selection committee of ten. Each member of the committee gives an initial ranking of the applicants and no one on the committee sees either the ranking by the HR professionals or the initial rankings of the other committee members. Ties are not permitted in any ranking. It is of interest to know if the rankings of the HR professionals and the initial rankings of the selection committee are correlated.

The scores for both the treatment (HR applicant ranking) and response (selection committee member ranking) are both 1, 2, 3, 4, and 5. Thus, $S_{XXj} = S_{YYj} = 10$ for all j. Routine calculations give $S_{XYj} = 10, 1, 2, 10, 1, 2, 10, 1, -3, 6$ for $j = 1, \ldots, 10$, respectively. It follows that $\sum_{j=1}^{b} S_{XYj} = 40$, $\sum_{j=1}^{b} S_{XXj} S_{YYj} = 1000$, and since $n_{\bullet\bullet j} = 5$ for $j = 1, \ldots, 10$, $S_C = 6.4$. This has χ_1^2 p-value 0.011. At the 0.05 level, the correlation can be regarded as non-zero. There appears to be good agreement between the rankings given by the HR personnel and the selection committee. As a check, the error sum of squares for the one-way ANOVA with blocks (committee member) as treatment is 100, as it should be.

The rank sums are 17, 32, 31, 34, and 36.

Since the stratum sums of squares and sums of products are required to calculate S_C, it is very little additional calculation to calculate the Pearson correlations and stratum CMH C statistics, S_{Cj}, say, to assess linear association on each stratum. The Pearson correlations for each selection committee member are given in the Table 4.5. The distribution of $(n - 1)\, r_P^2$ is asymptotically χ_1^2 and as the stratum p-values for each committee member, also given in Table 4.5, are based on only five observations, they should be regarded as a guide only. However, even by the values of the Pearson correlations alone, the main contributors to the significant overall correlation are committee members one, four, and seven.

Table 4.5 Rankings of five applicants by a selection committee.

	Committee member									
Applicant	1	2	3	4	5	6	7	8	9	10
A	1	2	2	1	2	2	1	2	2	2
B	2	5	3	2	5	3	2	5	4	1
C	3	1	5	3	1	5	3	1	5	4
D	4	4	1	4	4	1	4	4	3	5
E	5	3	4	5	3	4	5	3	1	3
r_P	1	0.1	0.2	1	0.1	0.2	1	0.1	−0.3	0.6
r_P p-values	0.046	0.842	0.689	0.046	0.842	0.689	0.046	0.842	0.549	0.230
S_{XYj}	10	1	2	10	1	2	10	1	−3	6

```
a_ij = matrix(1:5, ncol = 10, nrow = 5)
b_hj = matrix(1:5, ncol = 10, nrow = 5)
CMH(treatment = applicant, response = applicant_ranking,
     strata = committee_member, a_ij = a_ij, b_hj = b_hj,
     test_OPA = FALSE, test_GA = FALSE, test_MS = FALSE)
```

```
##
##            Cochran Mantel Haenszel Tests
##
##                    S df p-value
## Correlation 6.4    1 0.01141
##
##
## Correlations and statistics by strata:
##       S_C p-value  r_P S_XX S_XY S_YY
## 1    4.00  0.0455  1.0   10   10   10
## 2    0.04  0.8415  0.1   10    1   10
## 3    0.16  0.6892  0.2   10    2   10
## 4    4.00  0.0455  1.0   10   10   10
## 5    0.04  0.8415  0.1   10    1   10
## 6    0.16  0.6892  0.2   10    2   10
## 7    4.00  0.0455  1.0   10   10   10
## 8    0.04  0.8415  0.1   10    1   10
## 9    0.36  0.5485 -0.3   10   -3   10
## 10   1.44  0.2301  0.6   10    6   10
```

4.5 The Friedman Tests are CMH MS Tests

Here it will be shown that even when there are ties, if the ranks are used as scores for the CMH MS test, the adjusted Friedman statistic Fdm_A is equal to W_{MS}.

We begin by making the notations compatible. Define the indicator variable N_{ihj} which equals 1 if an observation on block j is of treatment i and is categorised into the hth category with score b_{hj}, and zero otherwise. Then $b_{hj}N_{ihj} = r_{ij}$, the rank given to treatment i on block j. We assume that on every block $\sum_{h=1}^{c} b_{hj}N_{\bullet hj} = \sum_{i=1}^{t} r_{ij} = \frac{t(t+1)}{2}$: the assignment of ranks on every block is such that the rank sum is the same as it would be for distinct ranks.

It was shown in Section 4.4.3 that

$$W_{MS} = \frac{t\sum_{i=1}^{t} V_i^2}{\sum_{j=1}^{b} S_j^2} \text{ in which } (t-1)S_j^2 = t\sum_{h=1}^{c} b_{hj}^2 n_{\bullet hj} - \left(\sum_{h=1}^{c} b_{hj} n_{\bullet hj}\right)^2.$$

We first address the numerator, then the denominator.

On the jth block there is just one response for every treatment, so $M_{ij} = r_{ij}$ and $E[M_{ij}] = \sum_{h=1}^{c} \frac{b_{hj} n_{\bullet hj}}{t} = \frac{t+1}{2}$. Thus, $V_i = M_{i\bullet} - E[M_{i\bullet}] = R_i - \frac{b(t+1)}{2}$ and

$$\sum_{i=1}^{t} V_i^2 = \sum_{i=1}^{t} \left[R_i - \frac{b(t+1)}{2} \right]^2 = \sum_{i=1}^{t} R_i^2 - \frac{b^2 t(t+1)^2}{4}.$$

To simplify the S_j^2, we need, first, $\sum_{h=1}^{c} b_{hj} n_{\bullet hj}$, that we have already assumed to be $\frac{t(t+1)}{2}$ for all j, and second, the $\sum_{h=1}^{c} b_{hj}^2 n_{\bullet hj}$. However, $\sum_{h=1}^{c} b_{hj}^2 N_{ihj} = r_{ij}^2$, so that $\sum_{h=1}^{c} b_{hj}^2 N_{\bullet h\bullet} = \sum_{i=1}^{t} r_{ij}^2$. Hence,

$$(t-1) S_j^2 = t \sum_{h=1}^{c} b_{hj}^2 n_{\bullet hj} - \left(\sum_{h=1}^{c} b_{hj} n_{\bullet hj} \right)^2 = t \sum_{i=1}^{t} r_{ij}^2 - \left[\frac{t(t+1)}{2} \right]^2$$

and

$$(t-1) \sum_{j=1}^{b} S_j^2 = t \sum_{i=1}^{t} \sum_{j=1}^{b} r_{ij}^2 - b \left[\frac{t(t+1)}{2} \right]^2 = t \left[\sum_{i=1}^{t} \sum_{j=1}^{b} r_{ij}^2 - \frac{bt(t+1)^2}{4} \right].$$

Substituting gives

$$W_{MS} = \frac{t \sum_{i=1}^{t} V_i^2}{\sum_{j=1}^{b} S_j^2} = \frac{(t-1) \left[\sum_{i=1}^{t} R_i^2 - b^2 \frac{t(t+1)^2}{4} \right]}{\sum_{i=1}^{t} \sum_{j=1}^{b} r_{ij}^2 - \frac{bt(t+1)^2}{4}} = Fdm_A.$$

One consequence of these results is that since the null hypothesis for the CMH MS test is equality of treatment means, the null hypothesis for the Friedman tests is equality of treatment mean ranks.

4.6 Relating the CMH MS and ANOVA F Tests

We now show that the CMH MS test statistic is simply related to the ANOVA F test statistic. This means that the statistics are just giving different representations of the same test. Since the ANOVA F test is clearly a location test, so is the CMH MS test, and since the Friedman tests are CMH MS tests, they are also location tests. This tightens our earlier statement that the conventional wisdom is that the Friedman test assesses whether or not the treatment distributions are consistent, but the test is known to be sensitive to differences in the means of the treatment ranks. In fact, the Friedman tests are tests of equality of mean treatment ranks.

Recall again that for the RBD $n_{i\bullet j} = 1$, $n_{\bullet\bullet j} = t$, $n_{i\bullet\bullet} = b$, $n_{\bullet\bullet\bullet} = bt$. Now, consider $\sum_{h=1}^{c} b_{hj} N_{ihj} = x_{ij}$, say. Note that $x_{\bullet j} = \sum_{h=1}^{c} b_{hj} N_{\bullet hj}$. On the jth block, there is only one observation of treatment i, so for only one value of h is $n_{ihj} = 1$; otherwise, $n_{ihj} = 0$. Thus, for example, $x_{ij}^2 = \sum_{h=1}^{c} b_{hj}^2 N_{ihj}$.

Consider the data set $\{x_{ij}\}$. For the *two-way* ANOVA of these data, write SS_{Treat} and SS_{Error} for the factor (treatments) and error sums of squares, respectively, and MS_{Treat} and MS_{Error} for the corresponding mean squares.

If the data $\{x_{ij}\}$ are analysed as a *one-way* ANOVA or CRD with blocks as treatments, then the error sum of squares is $SS_{\text{Treat}} + SS_{\text{Error}}$ and is given by

$$SS_{\text{Treat}} + SS_{\text{Error}} = \sum_{i=1}^{t}\sum_{j=1}^{b}\left(x_{ij} - \frac{x_{\bullet j}}{t}\right)^2 = \sum_{i=1}^{t}\sum_{j=1}^{b}x_{ij}^2 - \sum_{j=1}^{b}\frac{x_{\bullet j}^2}{t}$$

$$= \sum_{j=1}^{b}\left[\sum_{i=1}^{t}\left(\sum_{h=1}^{c}b_{hj}^2 N_{ihj}\right) - \frac{\left(\sum_{h=1}^{c}b_{hj}n_{\bullet hj}\right)^2}{t}\right]$$

$$= \sum_{j=1}^{b}\left[\sum_{h=1}^{c}b_{hj}^2 n_{\bullet hj} - \frac{\left(\sum_{h=1}^{c}b_{hj}n_{\bullet hj}\right)^2}{t}\right]$$

$$= (t-1)\sum_{j=1}^{b}\frac{S_j^2}{t}.$$

For the data set $\{x_{ij}\}$, the two-way ANOVA has treatment sum of squares (note that $\bar{x}$ means the mean of $\{x_i\}$, while $x_\bullet$ means the sum of the $\{x_i\}$)

$$SS_{\text{Treat}} = \sum_{i=1}^{t}\sum_{j=1}^{b}(\bar{x}_{i\bullet} - \bar{x}_{\bullet\bullet})^2 = b\sum_{i=1}^{t}\left[\frac{x_{i\bullet}}{b} - \frac{x_{\bullet\bullet}}{bt}\right]^2$$

$$= \frac{1}{b}\sum_{i=1}^{t}\left[\sum_{h=1}^{c}\sum_{j=1}^{b}b_{hj}N_{ihj} - \sum_{h=1}^{c}\sum_{j=1}^{b}\frac{b_{hj}n_{\bullet hj}}{t}\right]^2$$

$$= \frac{1}{b}\sum_{i=1}^{t}V_i^2.$$

It follows that

$$W_{\text{MS}} = \frac{t\sum_{i=1}^{t}V_i^2}{\sum_{j=1}^{b}S_j^2} = \frac{btSS_{\text{Treat}}}{\frac{t(SS_{\text{Treat}}+SS_{\text{Error}})}{t-1}}$$

$$= \frac{b(t-1)F}{b-1+F},$$

on using $MS_{\text{Treat}} = \frac{SS_{\text{Treat}}}{t-1}$, $MS_{\text{Error}} = \frac{SS_{\text{Error}}}{(b-1)(t-1)}$ and $F = \frac{MS_{\text{Treat}}}{MS_{\text{Error}}}$, the F test statistic for treatments in the two-way ANOVA. Equivalently,

$$F = \frac{(b-1)W_{\text{MS}}}{b(t-1) - W_{\text{MS}}}.$$

The significance of this result is twofold. First, calculations can be done using standard ANOVA routines, such as those available from click and point software. Second, *p*-values can be found using either W_{MS} and its asymptotic χ^2_{t-1} distribution or F and its $F_{t-1, (b-1)(t-1)}$ distribution. The empirical study in Section 4.7 is intended to give guidance on which is preferable. In the examples, we have analysed there has often been little difference between these methods.

The ANOVA F statistic is in fact the *rank transform statistic* for this design. As with the CRD, the development here shows that the rank transform statistic is a CMH MS statistic, although this is not the case for other designs.

4.6.1 Jams Data Revisited

We previously found that the unadjusted and adjusted Friedman tests have χ^2_2 *p*-values of 0.216 and 0.152, respectively. Using the relationship earlier with the adjusted Friedman statistic, the ANOVA F statistic F is 2.1572 with $F_{2,7}$ *p*-value 0.153. This is confirmed by direct calculation.

```
friedman(y = sweetness, groups = type, blocks = judge)

##
##          Friedman Rank Sum Test
##
## Test statistic is adjusted for ties
## data:   sweetness type judge
## Fdm_A = 3.769231, df = 2, p-value = 0.1518875
##
## For the group levels A B C,
## the rank sums are:   12.5 19.5 16

(F_stat = (8-1)*3.769/(8*(3-1)-3.769))

## [1] 2.15706

pf(q = F_stat, df1 = 3-1, df2 = (8-1)*(3-1), lower.tail = F)

## [1] 0.1525436
```

4.7 Simulation Study

We have established that the Friedman test and ANOVA F test using ranks are equivalent in the sense that if the critical values are chosen appropriately, both will lead to the same conclusion. However, the sampling distributions are different, and an important question is, which more closely approximates an intended 0.05 significance level: the Friedman using the asymptotic χ^2 distribution, or the ANOVA F test using the F distribution? It is generally accepted that the F test is preferable when there are no ties, but a new study is needed to assess the situation when there are ties. The treatment here is parallel to that in Section 3.8 for the CRD.

Consider a RBD with t treatments and b blocks. Take a grid with $t = 3, 7, 12$, and 20 and $b = 4, 7$, and 10. This gives sample sizes between 12 and 200. The upper end of this range should be sufficient for the asymptotic χ^2 distribution for the Friedman statistic to be viable.

A random sample was taken by first generating bt uniform $(0, 1)$ values and categorising these into c intervals of equal lengths. The categorised values were then ranked, assigning mid-ranks to ties, and randomly allocated consistent with the design. Then Fdm_A was calculated, and the null hypothesis of no treatment effects was rejected for values greater than the critical point of the χ^2_{t-1} distribution. Similarly, $F = \frac{(b-1)Fdm_A}{b(t-1-Fdm_A)}$ was calculated and the null hypothesis rejected for values greater than the critical point of the $F_{t-1, n-t}$ distribution. The proportion of rejections in 1,000,000 samples with nominal significance level 0.05 was recorded. The resulting standard error of the estimates is 0.0002.

We explored the effect of different values of c and found little change between small and large values. We give results for $c = 3, 10, 20, 50$, and 100, representing heavy categorisation becoming increasingly lighter.

For heavy categorisation and small sample sizes some values of zero occurred for C_{RBD}, so that Fdm_A and F could not be calculated. These reflect samples in which all blocks were uninformative: on each block, all observations were given the same rank. Such samples were discarded and replaced.

Our naïve expectation was that the χ^2 approach would be superior for larger values of b, since the asymptotic distribution requires b to approach infinity. We expected F to do well for larger c, on the basis that if there are too few categories the F statistic will have fewer achievable values and hence, the data couldn't be considered to be consistent with normality. This would be balanced by the well-known robustness of the F test.

The F proportions were almost always closer to 0.05 than the χ^2 proportions, with the majority of exceptions being for t small. The F test sizes were almost always greater than 0.05, while those for χ^2 were almost always less. For a fixed number of blocks as the number of treatments increases (columns in Table 4.6) the χ^2 proportions almost always reduced, away from 0.05, while the F proportions reduced towards 0.05. For a fixed number of treatments both approaches almost always improve as the number of blocks increases (rows in Table 4.6). There was a clear stability in the test sizes for varying values c, especially for larger values of bt.

Table 4.6 Proportion of rejections using the Friedman and ANOVA F tests for a nominal significance level of 0.05 based on 1,000,000 simulations.

$c = 3$ categories		Number of blocks (b)		
		4	7	10
Number of	3	0.0323/0.0600	0.0455/0.0549	0.0461/0.0538
treatments (t)	7	0.0277/0.0568	0.0388/0.0535	0.0427/0.0524
	12	0.0270/0.0537	0.0376/0.0517	0.0417/0.0512
	20	0.0270/0.0520	0.0375/0.0510	0.0412/0.0506

$c = 10$ categories		Number of blocks (b)		
		4	7	10
Number of	3	0.0430/0.0777	0.0437/0.0577	0.0465/0.0557
treatments (t)	7	0.0277/0.0570	0.0390/0.0538	0.0427/0.0526
	12	0.0268/0.0538	0.0381/0.0522	0.0421/0.0517
	20	0.0270/0.0522	0.0375/0.0514	0.0415/0.0507

$c = 20$ categories		Number of blocks (b)		
		4	7	10
Number of	3	0.0538/0.0740	0.0445/0.0563	0.0461/0.0545
treatments (t)	7	0.0280/0.0574	0.0390/0.0537	0.0427/0.0526
	12	0.0274/0.0540	0.0375/0.0514	0.0417/0.0513
	20	0.0270/0.0520	0.0374/0.0511	0.0415/0.0510

$c = 50$ categories		Number of blocks (b)		
		4	7	10
Number of	3	0.0620/0.0707	0.0476/0.0542	0.0461/0.0504
treatments (t)	7	0.0279/0.0571	0.0396/0.0542	0.0424/0.0522
	12	0.0268/0.0534	0.0381/0.0523	0.0424/0.0519
	20	0.0271/0.0525	0.0374/0.0512	0.0412/0.0506

$c = 100$ categories		Number of blocks (b)		
		4	7	10
Number of	3	0.0656/0.0700	0.0497/0.0535	0.0459/0.0481
treatments (t)	7	0.0280/0.0570	0.0392/0.0538	0.0429/0.0527
	12	0.0270/0.0538	0.0378/0.0519	0.0418/0.0513
	20	0.0273/0.0521	0.0374/0.0512	0.0415/0.0509

The first entry in each cell uses the χ^2 sampling distribution, while the second uses the F sampling distribution.

Our set-up reflects how data are obtained when judges (blocks) assign scores to treatments, as in a Likert study, and for each judge scores are ranked. We acknowledge that other set-ups are possible.

4.8 Wald Test Statistics in the RBD

The development in this section parallels that in Section 3.9, so certain aspects will be glossed over. A model is developed, and from that model, a Wald-type test statistic is derived. This requires MLEs that are given; their variances and covariances are derived using the extended hypergeometric distribution.

As previously suppose that for $i = 1, \ldots, t$, $j = 1, \ldots, b$, and $h = 1, \ldots, c$, $N_{ihj} = 1$ if treatment i is assessed as being in the hth category on block j, and zero otherwise. Put

$$P\left(N_{ihj} = 1\right) = \theta_{ih} + \beta_{hj},$$

where the θ_{ih} reflects the inclination for treatment i to be categorised in the hth category over and above the consensus of the other treatments, and where β_{hj} is the probability, averaging over treatments, that a treatment on block j is categorised into the hth category. These parameters mirror treatment and block parameters in the traditional ANOVA model.

Put $\theta_i = \left(\theta_{i1}, \ldots, \theta_{ic}\right)^{\mathrm{T}}$, $i = 1, \ldots, t$ and $\theta^{\mathrm{T}} = \left(\theta_1, \ldots, \theta_t\right)$. To test the null hypothesis H : $\theta = 0$ against K : $\theta \neq 0$ a Wald statistic of the form $\hat{\theta}^{\mathrm{T}} \mathrm{Cov}^{-}\left(\hat{\theta}\right) \hat{\theta}$ may be constructed.

To develop appropriate tests of H against K, we first need to consider constraints on the defined parameters and then develop estimators of them. By the definition of β_{hj}, $\sum_{i=1}^{t} P\left(N_{ihj} = 1\right) = \sum_{i=1}^{t}\left(\theta_{ih} + \beta_{hj}\right) = t\beta_{hj}$. Thus, $\theta_{1h} + \cdots + \theta_{th} = \theta_{\bullet h} = 0$ for all h. On block j treatment i is categorised into one of the categories, so $\sum_{h=1}^{c} P\left(N_{ihj} = 1\right) = \sum_{h=1}^{c}\left(\theta_{ih} + \beta_{hj}\right) = \theta_{i\bullet} + \beta_{\bullet j} = 1$. This must be true when there is no treatment effect, so $\beta_{\bullet j} = 1$ for all j. These last two results show that $\theta_{i\bullet} = 0$ for all i. There are thus $(c-1)(t-1)$ algebraically independent θ_{ih} and $b(c-1)$ algebraically independent β_{hj}.

By definition $\beta_{hj} = \frac{n_{\bullet hj}}{t}$ for $h = 1, \ldots, t$. Equating the sum over blocks of the (i, h)th outcomes, $\sum_{j=1}^{b} N_{ihj} = N_{ih\bullet}$ with its expectation, $\mathrm{E}\left[\sum_{j=1}^{b} N_{ihj}\right] = \sum_{j=1}^{b}\left(\theta_{ih} + \beta_{hj}\right) = b\theta_{ih} + \beta_{h\bullet}$ gives $N_{ih\bullet} = b\hat{\theta}_{ih} + \frac{n_{\bullet h\bullet}}{t}$, so that $b\hat{\theta}_{ih} = N_{ih\bullet} - \frac{n_{\bullet h\bullet}}{t}$.

This treatment identifies the $\hat{\theta}_{ih}$ as method of moments estimators. In fact, they are also maximum likelihood estimators. To see this, note that the estimators of the cell probabilities are the cell proportions. The collection of θs and βs are 1 - 1 functions of the cell probabilities, and so the given estimators of the θs are also maximum likelihood estimators. The cell probabilities are bounded between zero and one, and this implies bounds on the θs and βs. However, it isn't necessary to investigate these.

To evaluate the Wald-type test statistic, we need to develop the variances and covariances of the $\hat{\theta}_{ih}$. Using standard distribution theory for the extended hypergeometric distribution, the covariance between N_{ijh} and $N_{i'h'j}$ is

$$\text{Cov}\left(N_{ijh}, N_{i'h'j}\right) = \frac{n_{i\bullet j} n_{\bullet hj} \left(\delta_{ii'} n_{\bullet\bullet j} - n_{i'\bullet j}\right) \left(\delta_{hh'} n_{\bullet\bullet j} - n_{\bullet h'j}\right)}{n_{\bullet\bullet j}^2 \left(n_{\bullet\bullet j} - 1\right)}.$$

See, for example, Landis et al. (1979).

Now for the RBD

$$\text{Cov}\left(N_{ijh}, N_{i'h'j}\right) = \frac{\left(\delta_{ii'} t - 1\right) \beta_{hj} \left(\delta_{hh'} - \beta_{h'j}\right)}{t - 1}.$$

Henceforth, it is assumed that the null hypothesis of no treatment effects is true, and that both $i \neq i'$ and $h \neq h'$. Using the early moments of the extended hypergeometric distribution,

$$\text{E}\left[N_{ihj}\right] = \beta_{hj},$$
$$\text{Var}\left(N_{ihj}\right) = \beta_{hj}\left(1 - \beta_{hj}\right),$$
$$\text{Cov}\left(N_{ijh}, N_{i'hj}\right) = -\frac{\beta_{hj}(1 - \beta_{hj})}{t - 1},$$
$$\text{Cov}\left(N_{ijh}, N_{ih'j}\right) = -\beta_{hj}\beta_{h'j}, \text{ and}$$
$$\text{Cov}\left(N_{ijh}, N_{i'h'j}\right) = \frac{\beta_{hj}\beta_{h'j}}{t - 1}.$$

Then $\text{E}\left[N_{ih\bullet}\right] = \frac{n_{\bullet h\bullet}}{t}$, and, writing $\sigma_{hh} = \sum_{j=1}^{b} \beta_{hj}\left(1 - \beta_{hj}\right)$ and $\sigma_{hh'} = -\sum_{j=1}^{b} \beta_{hj}\beta_{h'j}$,

$$\text{Var}\left(N_{ih\bullet}\right) = \sigma_{hh}, \quad \text{Cov}\left(N_{ih\bullet}, N_{i'h\bullet}\right) = -\frac{\sigma_{hh}}{t - 1},$$
$$\text{Cov}\left(N_{ih\bullet}, N_{ih'\bullet}\right) = \sigma_{hh'} \text{ and } \text{Cov}\left(N_{ih\bullet}, N_{i'h'\bullet}\right) = -\frac{\sigma_{hh'}}{t - 1}.$$

For example,

$$\text{Cov}\left(N_{ih\bullet}, N_{ih'\bullet}\right) = \text{Cov}\left(\sum_{j=1}^{b} N_{ihj}, \sum_{j=1}^{b} N_{ih'j}\right) = \sum_{j=1}^{b} \text{Cov}\left(N_{ihj}, N_{ih'j}\right)$$

$$= -\sum_{j=1}^{b} \beta_{hj}\beta_{h'j} = \sigma_{hh'}.$$

Using these results and rules for the variances and covariances of linear combinations of random variables,

$$b^2 \operatorname{Var}\left(\hat{\theta}_{ih}\right) = \sigma_{hh}, \quad b^2 \operatorname{Cov}\left(\hat{\theta}_{ih}, \hat{\theta}_{i'h}\right) = -\frac{\sigma_{hh}}{t-1},$$

$$b^2 \operatorname{Cov}\left(\hat{\theta}_{ih}, \hat{\theta}_{ih'}\right) = \sigma_{hh'} \text{ and } b^2 \operatorname{Cov}\left(\hat{\theta}_{ih}, \hat{\theta}_{i'h'}\right) = -\frac{\sigma_{hh'}}{t-1}.$$

For example,

$$(bt)^2 \operatorname{Var}\left(\hat{\theta}_{ih}\right) = (t-1)^2 \operatorname{Var}\left(N_{ih\bullet}\right) + \sum_{\substack{u=1 \\ u\neq i}}^{t} \operatorname{Var}\left(N_{uh\bullet}\right)$$

$$- 2(t-1) \sum_{\substack{u=1 \\ u\neq i}}^{t} \operatorname{Cov}\left(N_{ih\bullet}, N_{uh\bullet}\right)$$

$$+ \sum_{\substack{u=1 \\ }}^{t} \sum_{\substack{v=1 \\ v\neq i}}^{t} \operatorname{Cov}\left(N_{uh\bullet}, N_{vh\bullet}\right)$$

$$= (t-1)^2 \sigma_{hh} + (t-1)\sigma_{hh} - 2(t-1)^2 \left(-\frac{\sigma_{hh}}{t-1}\right)$$

$$+ (t-1)(t-2)\left(-\frac{\sigma_{hh}}{t-1}\right) = t\sigma_{hh}$$

and

$$(bt)^2 \operatorname{Cov}\left(\hat{\theta}_{ih}, \hat{\theta}_{ih'}\right)$$

$$= \operatorname{Cov}\left(N_{ih\bullet}(t-1) - \sum_{\substack{u=1 \\ u\neq i}}^{t} N_{uh\bullet}, N_{ih'\bullet}(t-1) - \sum_{\substack{v=1 \\ v\neq i}}^{t} N_{vh'\bullet}\right)$$

$$= (t-1)^2 \operatorname{Cov}\left(N_{ih\bullet}, N_{ih'\bullet}\right) + \operatorname{Cov}\left(\sum_{\substack{u=1 \\ u\neq i}}^{t} N_{uh\bullet}, \sum_{\substack{v=1 \\ v\neq i}}^{t} N_{vh'\bullet}\right)$$

$$- (t-1) \operatorname{Cov}\left(\sum_{\substack{u=1 \\ u\neq i}}^{t} N_{uh\bullet}, N_{ih'\bullet}\right) - (t-1) \operatorname{Cov}\left(N_{ih\bullet}, \sum_{\substack{v=1 \\ v\neq i}}^{t} N_{vh'\bullet}\right)$$

$$= t^2 \sigma_{hh'}$$

ultimately.

Recall that $\theta_i = \left(\theta_{i1}, \ldots, \theta_{ic}\right)^{\mathrm{T}}$, $i = 1, \ldots, t$, $\theta^{\mathrm{T}} = \left(\theta_1^{\mathrm{T}}, \ldots, \theta_t^{\mathrm{T}}\right)$ and write $\Sigma = \{\sigma_{hk}\} = \operatorname{diag}\left(\sum_{j=1}^{b} \beta_{hj}\right) - \sum_{j=1}^{b} \beta_{hj}\beta_{kj}$. The diagonal elements of $b^2 \operatorname{Cov}\left(\hat{\theta}_i\right)$ are σ_{hh}, and the off-diagonal elements are $\sigma_{hh'}$. So this matrix is Σ. The diagonal elements of $b^2 \operatorname{Cov}\left(\hat{\theta}_i, \hat{\theta}_{i'}\right)$ are $-\frac{\sigma_{hh}}{t-1} = \frac{t}{t-1}\left(-\frac{\sigma_{hh}}{t}\right)$ and the off-diagonal elements are $\frac{t}{t-1}\left(-\frac{\sigma_{hh'}}{t}\right)$. So this matrix is $-\frac{1}{t-1}\Sigma = \frac{t}{t-1}\left(-\frac{1}{t}\right)\Sigma$. If $\otimes$ is the Direct or Kronecker product, then

$$b^2\,\mathrm{Cov}\left(\hat{\theta}\right) = \frac{t}{t-1}\left(\left[I_t - \frac{1_t 1_t^T}{t}\right]\otimes\Sigma\right),$$

from which the Moore–Penrose inverse of $\mathrm{Cov}\left(\hat{\theta}\right)$ is

$$\mathrm{Cov}^-\left(\hat{\theta}\right) = b^2\frac{t-1}{t}\left(I_t - \frac{1_t 1_t^T}{t}\right)\otimes\Sigma^{-1}.$$

This uses the facts that $\frac{1_t 1_t^T}{t}$ and hence, $I_t - \frac{1_t 1_t^T}{t}$ are idempotent, and that the Moore–Penrose inverse of an idempotent matrix is itself. Now, $1_t^T\hat{\theta}_i = 0$ because $\theta_{i\bullet} = 0$ for all i and so the Wald-type test statistic is W_{GA}, given by

$$W_{GA} = \hat{\theta}^T\mathrm{Cov}^-\left(\hat{\theta}\right)\hat{\theta} = b^2\frac{t-1}{t}\sum_{i=1}^{t}\hat{\theta}_i\hat{\Sigma}^{-1}\hat{\theta}_i.$$

The CMH GA test statistic S_{GA} is a quadratic form with vector $U_\bullet - \mathrm{E}\left[U_\bullet\right] = N_{ih\bullet} - \mathrm{E}\left[N_{ih\bullet}\right]$, while the elements of the vector defining W_{GA} are $\hat{\theta}_{ih} = \frac{N_{ih\bullet} - \frac{N_{\bullet h\bullet}}{t}}{b}$. These differ by only a constant. In both cases, the matrices in the quadratic forms are generalised inverses of the covariance matrices. Thus, $S_{GA} = W_{GA}$.

By the central limit theorem (CLT) $\hat{\theta}$ is asymptotically normal. The quadratic form W_{GA} has asymptotic chi-squared distribution with degrees of freedom those of $\mathrm{Cov}\left(\hat{\theta}\right)$. These are the product of the degrees of freedom associated with the factor matrices in $\mathrm{Cov}\left(\hat{\theta}\right)$, namely, $t-1$ in $I_t - \frac{1_t 1_t^T}{t}$ and $c-1$ in $\hat{\Sigma}$. Thus, W_{GA} has an asymptotic $\chi^2_{(c-1)(t-1)}$ distribution.

For the CMH MS and C tests, it is not clear how to establish that they are Wald tests unless it is also assumed that there are no ties and scores are not block dependent, which we now do.

Recall the RBD constraints $n_{i\bullet j} = 1$ for all i and j, which imply $n_{\bullet\bullet j} = t$, $n_{i\bullet\bullet} = b$, $n_{\bullet\bullet\bullet} = bt$ and $p_{i\bullet j} = \frac{n_{i\bullet\bullet}}{n_{\bullet\bullet\bullet}} = \frac{1}{t}$. Now, if there are no ties, then $n_{\bullet hj} = 1$: the hth rank is assigned just once on any block. Thus, on block j, $\mathrm{E}\left[N_{ihj}\right] = \frac{n_{i\bullet j}n_{\bullet hj}}{n_{\bullet\bullet j}} = \frac{1}{t}$. Moreover, if the scores are not block dependent, then $b_{hj} = b_h$ and $\sum_{h=1}^{c}\sum_{j=1}^{b}b_{hj}N_{ihj} = \sum_{h=1}^{c}b_h N_{ih\bullet}$ with expectation $\sum_{h=1}^{c}\sum_{j=1}^{b}\frac{b_h n_{\bullet hj}}{t} = b\sum_{h=1}^{c}\frac{b_h}{t} = \frac{b(t+1)}{2}$.

As previously, assume $\mathrm{P}\left(N_{ihj} = 1\right) = \theta_{ih} + \beta_{hj}$, and deduce that $b\theta_{ih} = N_{ih\bullet} - \frac{b}{t}$ so that $\Theta = \{\theta_{ih}\}$ has maximum likelihood estimator $\hat{\Theta} = \{\hat{\theta}_{ih}\}$ in which $\hat{\theta}_{ih} = \frac{N_{ih\bullet}}{b} - \frac{1}{t}$. Write $s_b = \{b_h\}$ for the vector of block

scores. Then $\hat{\Theta} s_b$ has elements $\sum_{h=1}^{c} b_h \left(\frac{N_{ih\bullet}}{b} - \frac{1}{t} \right) = \frac{M_{i\bullet}}{b} - \frac{t+1}{2} = \frac{(M_{i\bullet} - \mathrm{E}[M_{i\bullet}])}{b}$.
This follows because $b_h = h$ and $\sum_{h=1}^{c} b_h = \frac{t(t+1)}{2}$.

Thus, $\hat{\Theta} s_b = \frac{M - \mathrm{E}[M]}{b}$. Since W_{MS} is a quadratic form based on $M - \mathrm{E}[M]$, it is also a quadratic form based on $\hat{\Theta} s_b$. It is thus the Wald-type test statistic for testing $\Theta s_b = 0$ against $\Theta s_b \neq 0$.

Now suppose that both sets of scores, $\{a_{ij}\}$ and $\{b_{hj}\}$, are not block dependent. Define the vector of treatment scores $s_a = \{a_i\}$ and the vector of block scores $s_b = \{b_h\}$. As before $\Theta = \{\theta_{ih}\}$ has maximum likelihood estimator $\hat{\Theta} = \{\hat{\theta}_{ih}\}$ in which $\hat{\theta}_{ih} = \frac{N_{ih\bullet}}{b} - \frac{1}{t}$. Then $s_a^{\mathrm{T}} \hat{\Theta} s_b$ has elements $\sum_{i=1}^{t} \sum_{h=1}^{c} a_i \left(\frac{N_{ih\bullet}}{b} - \frac{1}{t} \right) b_h$.

Recall that for the RBD the CMH C statistic is, using $\mathrm{E}\left[N_{ihj}\right] = \frac{1}{t}$, proportional to $\sum_{i=1}^{t} \sum_{h=1}^{c} \sum_{j=1}^{b} a_i \left(N_{ihj} - \frac{1}{t} \right) b_h = \sum_{i=1}^{t} \sum_{h=1}^{c} a_i \left(N_{ih\bullet} - \frac{b}{t} \right) b_h = b a \hat{\Theta} b$. This makes use of the facts that $n_{\bullet\bullet j} = t$, that

$$n_{\bullet\bullet j} a_j^{\mathrm{T}} V_{\mathrm{T}j} a_j = \sum_{i=1}^{t} a_{ij}^2 n_{i\bullet j} - \frac{\left(\sum_{i=1}^{t} a_{ij} n_{i\bullet j} \right)^2}{n_{\bullet\bullet j}} = \mathrm{Var}\left(\{a_i\} \right),$$

and similarly that

$$n_{\bullet\bullet j} b_j^{\mathrm{T}} V_{\mathrm{C}j} b_j = \sum_{h=1}^{c} b_{hj}^2 n_{\bullet h j} - \frac{\left(\sum_{h=1}^{c} b_{hj} n_{\bullet h j} \right)^2}{n_{\bullet\bullet j}} = \mathrm{Var}\left(\{b_h\} \right).$$

Since $W_{\mathrm{C}} = \frac{C^2}{\mathrm{Var}(C)}$ if $\phi = s_a^{\mathrm{T}} \hat{\Theta} s_b$ it is also $\frac{\hat{\phi}^2}{\hat{\mathrm{Var}}(\hat{\phi})}$. It is thus the Wald-type test statistic for testing $\phi = 0$ against $\phi \neq 0$.

All of the tests that have been shown to be Wald tests enjoy the asymptotic optimality properties of these tests, and that the asymptotic sampling distributions are chi-squared.

Bibliography

Landis, J. R., Cooper, M. M., Kennedy, T., and Koch, G. G. (1979). A computer program for testing average partial association in three-way contingency tables (PARCAT). *Computer Programs in Biomedicine*, 9(3):223–246.

Rayner, J. C. W. and Best, D. J. (2017). Unconditional analogues of Cochran-Mantel-Haenszel tests. *Australian & New Zealand Journal of Statistics*, 59(4):485–494.

Rayner, J. C. W. and Best, D. J. (2018). Extensions to the Cochran-Mantel-Haenszel mean scores and correlation tests. *Journal of Statistical Theory and Practice*, 12(3):561–574.

5

The Balanced Incomplete Block Design

5.1 Introduction

The Cochran–Mantel–Haenszel (CMH) tests apply when every treatment is present on every block or stratum and responses are categorical. These methods therefore don't apply to incomplete designs. Here we look at perhaps the most elementary of the incomplete designs, the balanced incomplete block design (BIBD), in part to see how CMH methods can aid analysis for such designs. However both the BIBD and the Latin square design (LSD) are of interest since, along with the completely randomised design (CRD) and the randomised block design (RBD), they comprise a quartet of elementary designs. We will find that although the CMH methodology doesn't directly apply to the BIBD and the LSD, the nonparametric ANOVA methodology that will be developed in later chapters does. See also Sections 8.8 and 10.8 for more on the BIBD.

5.2 The Durbin Tests

It may not always be possible for an experiment to have complete blocks. For example, in taste-testing palate fatigue may prevent tasters from tasting all treatments at a single sitting. One way to allow for this is to intentionally have incomplete blocks, but in a balanced format.

In the BIBD each of the b blocks contains k experimental units, each of the t treatments appears in r blocks, and every treatment appears with every other treatment precisely λ times. Necessarily

$$k < t, \quad r < b, \quad bk = rt, \quad \text{and} \quad \lambda(t-1) = r(k-1).$$

Supposed treatments are ranked within each block. If there are no ties then the null hypothesis of no treatment effects can be tested using Durbin's

An Introduction to Cochran–Mantel–Haenszel Testing and Nonparametric ANOVA,
First Edition. J.C.W. Rayner and G. C. Livingston Jr.
© 2023 John Wiley & Sons Ltd. Published 2023 by John Wiley & Sons Ltd.

statistic, D, where

$$D = \frac{12(t-1)}{rt(k^2-1)} \sum_{i=1}^{t} \left[R_i - \frac{r(k+1)}{2} \right]^2$$

$$= -\frac{3r(t-1)(k+1)}{k-1} + \frac{12(t-1)}{bk(k^2-1)} \sum_{i=1}^{t} R_i^2,$$

in which R_i is the sum of the ranks given to treatment i, $i = 1, \ldots, t$. Under the null hypothesis of no treatment effects, D asymptotically has the χ^2_{t-1} distribution.

If ties occur then an adjusted test statistic is given by

$$D_A = \frac{(t-1) \left[\sum_{i=1}^{t} R_i^2 - \frac{rbk(k+1)^2}{4} \right]}{\sum_{i=1}^{t} \sum_{j=1}^{b} r_{ij}^2 - \frac{bk(k+1)^2}{4}},$$

in which r_{ij} is the rank of treatment i on block j. If there are no ties then algebra similar to that in Section 3.3 shows that D_A reduces to D. As with the Kruskal–Wallis and Friedman tests, the basis for the adjustment is to replace in D the variance of the untied ranks with the variance of the tied ranks.

Provided rank sums are preserved, the test based on D_A earlier is applicable no matter how the ties are treated. If ties do occur and mid-ranks are used, then algebra similar to that in Section 3.3 may be used to show that $D_M = \frac{D}{C_{BIBD}}$ in which D_M is the Durbin test using mid-ranks and

$$C_{BIBD} = 1 - \sum_{g=1}^{G} \sum_{j=1}^{b} \frac{t_{gj}^3 - t_{gj}}{b(k-1)k(k+1)}.$$

As before, this form, a particular case of D_A, is convenient if there are few ties or if calculation is principally by hand.

5.3 The Relationship Between the Adjusted Durbin Statistic and the ANOVA F Statistic

Here we show that when the data are ranks adjusted for ties if they occur, the ANOVA F test statistic F is related to the adjusted Durbin statistic D_A by

$$F = \frac{edf\, D_A}{(t-1) \left[b(k-1) - D_A \right]},$$

in which $edf = bk - b - t + 1$, the error degrees of freedom. This is an algebraic relationship that assumes no model. When there are no ties $D_A = D$ and the relationship reduces to that derived, for example, in Rayner (2016, section 2.5.3). The relationship shows that the adjusted Durbin test and the ANOVA F test are essentially the same test, so both are tests for equality of treatment mean ranks, not merely for equality of treatment distributions. As has been shown in Chapters 3 and 4, parallel results hold for the adjusted Kruskal–Wallis and Friedman tests.

Aside: There are different possible ANOVA analyses for BIBDs. See, for example, Kuehl (2000, Section 8.5). Note that a BIBD can be regarded as a block design with missing data. Such designs are not *orthogonal* and so it is possible to calculate treatment sums of squares both adjusted and not adjusted for blocks. We use the former.

First note that as the block means are the same on every block, there are no block effects, and the block sum of squares is zero.

The observations are the r_{ij}, the ranks for treatment i on block j. The total sum of squares is

$$SS_{\text{Total}} = \sum_{i=1}^{t} \sum_{j=1}^{b} r_{ij}^2 - \frac{r_{\bullet\bullet}^2}{bk}$$

$$= \sum_{i=1}^{t} \sum_{j=1}^{b} r_{ij}^2 - \frac{bk(k+1)^2}{4},$$

since $r_{\bullet\bullet} = \frac{bk(k+1)}{2}$.

The adjusted treatment sum of squares is

$$SS_{\text{Treat}} = \frac{k}{\lambda t} \sum_{i=1}^{t}$$

$$\times \left[R_i - \left(\text{block average across blocks that contain treatment } i\right)\right]^2.$$

For any given i, there are r blocks containing treatment i and each block total is $\frac{k(k+1)}{2}$, so the block average across blocks that contain treatment i is $\frac{r(k+1)}{2}$. It follows that

$$SS_{\text{Treat}} = \frac{k}{\lambda t} \sum_{i=1}^{t} \left[R_i - \frac{r(k+1)}{2} \right]^2 = \frac{k(k+1)}{12} D.$$

This uses $\sum_{i=1}^{t} \left[R_i - \frac{r(k+1)}{2} \right]^2 = \frac{rt(k^2-1)D}{12(t-1)}$ from the equation for D and both $bk = rt$, and $\lambda(t-1) = r(k-1)$.

The error sum of squares is found by difference:

$$SS_{\text{Error}} = SS_{\text{Total}} - SS_{\text{Treat}} - SS_{\text{Block}}$$

$$= \sum_{i=1}^{t} \sum_{j=1}^{b} r_{ij}^2 - \frac{bk(k+1)^2}{4} - \frac{k}{\lambda t} \sum_{i=1}^{t} \left[R_i - \frac{r(k+1)}{2} \right]^2.$$

From the equation for D_A

$$\frac{D_A}{t-1} = \frac{\sum_{i=1}^{t} R_i^2 - \frac{rbk(k+1)^2}{4}}{\sum_{i=1}^{t} \sum_{j=1}^{b} r_{ij}^2 - \frac{bk(k+1)^2}{4}}.$$

The ANOVA F statistic is thus

$$
F = \frac{\frac{SS_{\text{Treat}}}{t-1}}{\frac{SS_{\text{Error}}}{\text{edf}}} = \frac{\text{edf}}{t-1}\left[\frac{\frac{k}{\lambda t}\sum_{i=1}^{t}\left[R_i - \frac{r(k+1)}{2}\right]^2}{\sum_{i=1}^{t}\sum_{j=1}^{b} r_{ij}^2 - \frac{bk(k+1)^2}{4} - \frac{k}{\lambda t}\sum_{i=1}^{t}\left[R_i - \frac{r(k+1)}{2}\right]^2} \right]
$$

$$
= \frac{\text{edf}}{t-1}\frac{\frac{\frac{k}{\lambda t}\sum_{i=1}^{t}\left[R_i - \frac{r(k+1)}{2}\right]^2}{\sum_{i=1}^{t}\sum_{j=1}^{b} r_{ij}^2 - \frac{bk(k+1)^2}{4}}}{1 - \frac{\frac{k}{\lambda t}\sum_{i=1}^{t}\left[R_i - \frac{r(k+1)}{2}\right]^2}{\sum_{i=1}^{t}\sum_{j=1}^{b} r_{ij}^2 - \frac{bk(k+1)^2}{4}}}
$$

$$
= \frac{\text{edf}}{t-1}\frac{\frac{\frac{k}{\lambda t}D_A}{t-1}}{1 - \frac{\frac{k}{\lambda t}D_A}{t-1}}
$$

$$
= \frac{\text{edf}}{t-1}\frac{D_A}{\frac{(t-1)\lambda t}{k} - D_A}
$$

$$
= \frac{\text{edf}\, D_A}{(t-1)\left[b(k-1) - D_A\right]}
$$

as previously indicated, since $bk(k-1) = (t-1)\lambda t$, which follows from the BIBD identities $\lambda(t-1) = r(k-1)$ and $bk = rt$.

As the observations here are ranks and are not normally distributed, only approximately does the ANOVA F statistic have the distribution $F_{t-1,\,bk-b-t+1}$.

It is well-known that the F test using the $F_{t-1,bk-b-t+1}$ distribution improves considerably on the Durbin test using the χ_{t-1}^2 distribution. See, for example, Best and Rayner (2014). However in Section 5.4 we expand on that study to better see the effect of categorisation.

5.3.1 Ice Cream Flavouring Data

Suppose we wish to compare seven vanilla ice-creams A, B, C, D, E, F, and G that are the same except for increasing amounts of vanilla flavouring. The data in the Table 5.1 were given in Conover (1998, p. 391) and analysed in Thas et al. (2012) using a BIBD. Each of the seven judges ranks without ties three ice creams. The rank sums are 8, 9, 4, 3, 5, 6, and 7.

A parametric analysis finds ice creams have a *p*-value of 0.0049. Judges have a *p*-value of 0.2210. It seems the ice cream means are significantly different at the 0.01 level. The Shapiro–Wilk test for normality of the residuals has *p*-value 0.4445. From this perspective the parametric model is acceptable. However treating the ranks as scores seems problematic; there are seven scores of 1, seven scores of 2, and seven scores of 3. This hardly seems to be normal, whatever the Shapiro–Wilk test may conclude.

For these data there are no ties and $b = t = 7$, $k = r = 3$. Since $\sum_{i=1}^{t} R_i^2 = 280$, $D = 12$. The χ_6^2 *p*-value is 0.0620. At the 0.05 level there is no evidence

Table 5.1 Scores for ice creams.

Judge	Variety						
	1	2	3	4	5	6	7
1	2	3		1			
2		3	1		2		
3			2	1		3	
4				1	2		3
5	3				1	2	
6		3				1	2
7	3		1				2
Variety sum	8	9	4	3	5	6	7

of a difference in the distributions of the ice creams. However using the formula earlier the F test statistic takes the value 8 with $F_{6,8}$ p-value 0.0049. From not quite significant at the 0.05 level using the χ^2_6 approximation, the F approximation results in significance at the 0.01 level.

```
lm_ice_cream = lm(as.numeric(rank)~variety + judge)
Anova(lm_ice_cream,type=3)

## Anova Table (Type III tests)
##
## Response: as.numeric(rank)
##               Sum Sq Df F value    Pr(>F)
## (Intercept) 10.6813  1 42.7253 0.000181 ***
## variety     12.0000  6  8.0000 0.004904 **
## judge        2.6667  6  1.7778 0.221005
## Residuals    2.0000  8
## ---
## Signif. codes:  0 '***' 0.001 '**' 0.01 '*' 0.05 '.' 0.1 ' ' 1

shapiro.test(lm_ice_cream$residuals)

##
##    Shapiro-Wilk normality test
##
## data:  lm_ice_cream$residuals
## W = 0.95628, p-value = 0.4445

durbin(y = rank, groups = variety, blocks = judge)

##
##          Durbin Rank Sum Test
##
## Test statistic is adjusted for ties
## data:  rank variety judge
```

```
## D_A = 12, df = 6, p-value = 0.0619688
##
## For the group levels 1 2 3 4 5 6 7,
## the rank sums are:  8 9 4 3 5 6 7

b = t = 7; k = r = 3
edf = b*k-b-t+1
(F_stat = edf*12/((t-1)*(b*(k-1)-12)))

## [1] 8

pf(q = F_stat, df1 = t-1, df2 = edf, lower.tail = F)

## [1] 0.004904419
```

5.3.2 Breakfast Cereal Data

The data in Table 5.2 are from Kutner et al. (2005, section 28.1). Each of five breakfast cereals is ranked by 10 judges, who each taste three cereals. Each cereal is assessed six times. Thus $t = 5, b = 10, k = 3$, and $r = 6$. The ANOVA F test on the ranks has F statistic 11.5556 with p-value 0.0001.

For these data $\sum_{i=1}^{t} \sum_{j=1}^{b} r_{ij}^2 = 137.5$ and $\sum_{i=1}^{t} R_i^2 = 785$, and using the direct formula $D_A = 14.8571$. Alternatively the uncorrected Durbin statistic

Table 5.2 Rankings for breakfast cereals.

			Cereal		
Judge	A	B	C	D	E
1	1.5	1.5	3		
2	1	2.5		2.5	
3	1.5	3			1.5
4	1		2	3	
5	1.5		3		1.5
6	1			3	2
7		2	3	1	
8		2.5	2.5		1
9		3		2	1
10			3	2	1
Sum	7.5	14.5	16.5	13.5	8

is $D = 13$ and $\sum_{g=1}^{G}\sum_{j=1}^{b}\left(t_{gj}^3 - t_{gj}\right) = 30$, since in five blocks there are two ties and in the remaining five blocks there are none. Thus $C_{\text{BIBD}} = \frac{7}{8}$ and $D_{\text{M}} = \frac{D}{C_{\text{BIBD}}} = 13 \times \frac{8}{7} = 14.8571$, as with D_{A}. The χ_4^2 p-value for D_{A} is 0.0050 whereas that for D is 0.0113. The values for F and D_{A} are consistent with the relationship given earlier in this section.

Best and Rayner (2014) also consider this example, but their value for D_{A} is in error.

```
Anova(lm(rank~type + judge), type = 3)

## Anova Table (Type III tests)
##
## Response: rank
##                 Sum Sq Df F value    Pr(>F)
## (Intercept)    1.8667  1   6.637 0.0202958 *
## type          13.0000  4  11.556 0.0001327 ***
## judge          2.1667  9   0.856 0.5796333
## Residuals      4.5000 16
## ---
## Signif. codes:  0 '***' 0.001 '**' 0.01 '*' 0.05 '.' 0.1 ' ' 1

b = 10; t = 5; k = 3; r = 6
durbin(y = rank, groups = type, blocks = judge)

##
##          Durbin Rank Sum Test
##
## Test statistic is adjusted for ties
## data:  rank type judge
## D_A = 14.85714, df = 4, p-value = 0.005006871
##
## For the group levels A B C D E,
## the rank sums are:   7.5 14.5 16.5 13.5 8

edf = b*k-b-t+1
(F_stat = edf*12/((t-1)*(b*(k-1)-12)))

## [1] 6

pf(q = F_stat, df1 = t-1, df2 = edf, lower.tail = F)

## [1] 0.003801088
```

5.4 Simulation Study

We have established that the Durbin test and ANOVA F test using ranks are equivalent in the sense that if the critical values are chosen appropriately, both will lead to the same conclusion. However the sampling distributions are different, and an important question is, which more closely approximates an intended 0.05 significance level: the Durbin test using the asymptotic χ^2 distribution, or the ANOVA F test using the F distribution?

Best and Rayner (2014) considered the BIBDs $(t, b, k, r) = (4, 6, 2, 3), (4, 4, 3, 3), (5, 10, 2, 4), (5, 5, 4, 4), (5, 10, 3, 6), (6, 15, 2, 5), (6, 10, 3, 5), (6, 15, 4, 10), (6, 20, 3, 10), (7, 7, 3, 3), (7, 7, 4, 4)$, and $(7, 21, 2, 6)$, so we do too. This gives sample sizes bk of between 12 and 60. At the upper end of this range this should be sufficient for the asymptotic χ^2 distribution for the Durbin statistic to be viable.

A random sample was taken by first generating bk uniform $(0, 1)$ values and categorising these into c intervals of equal lengths. Clearly $c = 3$ represents a severe categorisation, one that is unlikely to occur in practice. As c increases the categorisation becomes increasing less severe.

The categorised values were randomly allocated consistent with the design and then ranked within blocks, assigning mid-ranks to ties. Then both D and D_A were calculated, as was the proportion of values greater than the critical point of the χ^2_{t-1} distribution. Similarly $F = \frac{\text{edf } D_A}{(t-1)[b(k-1)-D_A]}$ was calculated as was the proportion of values greater than the critical point of the $F_{t-1,bk-b-t+1}$ distribution. In all cases the proportion of rejections in 1,000,000 samples with nominal significance level 0.05 was recorded. The resulting standard error of the estimates is 0.0002.

We emphasise that although the current study uses the same designs as Best and Rayner (2014) the treatment of ties is different; we do not use the more complicated approach in Brockhoff et al. (2004) that was implemented in Best and Rayner (2014). Moreover here we have 1,000,000 simulations instead of 100,000.

For heavy categorisation and small sample sizes, some values of zero occurred for C_{BIBD}. These reflect samples in which all blocks were uninformative: on each block all observations were given the same rank. Such samples were discarded and replaced.

The designs in Table 5.3 are arranged in increasing sample sizes bk. There is a clear improvement in all three tests as bk increases.

In terms of closeness to the nominal significance level uniformly the unadjusted Durbin test is worst, and the ANOVA F test is best.

Apart from the results for $c = 3$ there is a general, although not uniform, consistency across different values of c. Mostly there is an improvement as c increases.

Apart from the $(5, 10, 2, 4)$ design, results are satisfactory for the F test with bk at least 20. For the first three designs in Table 5.3 the two Durbin tests never reject the null hypothesis! The unadjusted Durbin test

Table 5.3 Proportion of rejections using the unadjusted Durbin, adjusted Durbin, and ANOVA F tests for a nominal significance level of 0.05 based on 1,000,000 simulations.

Design (t, b, k, r)	bk	Number of categories (c)				
		$c = 3$	$c = 10$	$c = 20$	$c = 50$	$c = 100$
(4, 6, 2, 3)	12	0.0000	0.0000	0.0197	0.0000	0.0000
		0.0032	0.0000	0.0000	0.0009	0.0000
		0.0000	0.0001	0.0000	0.0000	0.0000
(4, 4, 3, 3)	12	0.0000	0.0022	0.0798	0.0000	0.0000
		0.0896	0.0000	0.0000	0.0879	0.0000
		0.0000	0.0826	0.0000	0.0000	0.0798
(5, 10, 2, 4)	20	0.0000	0.0005	0.0554	0.0000	0.0000
		0.0995	0.0000	0.0000	0.1133	0.0000
		0.0000	0.1185	0.0000	0.0000	0.1198
(5, 5, 4, 4)	20	0.0024	0.0255	0.0612	0.0167	0.0269
		0.0639	0.0217	0.0271	0.0634	0.0242
		0.0264	0.0629	0.0253	0.0265	0.0634
(7, 7, 3, 3)	21	0.0000	0.0000	0.0615	0.0000	0.0003
		0.0673	0.0000	0.0001	0.0734	0.0000
		0.0000	0.0811	0.0000	0.0000	0.0843
(7, 7, 4, 4)	28	0.0014	0.0231	0.0583	0.0137	0.0234
		0.0589	0.0190	0.0244	0.0586	0.0225
		0.0247	0.0573	0.0236	0.0248	0.0568
(5, 10, 3, 6)	30	0.0043	0.0341	0.0575	0.0239	0.0358
		0.0585	0.0303	0.0361	0.0588	0.0338
		0.0360	0.0597	0.0339	0.0350	0.0602
(6, 15, 2, 5)	30	0.0002	0.0144	0.0638	0.0071	0.0186
		0.0589	0.0128	0.0201	0.0551	0.0178
		0.0210	0.0526	0.0199	0.0216	0.0521
(6, 10, 3, 5)	30	0.0018	0.0276	0.0582	0.0163	0.0288
		0.0611	0.0217	0.0282	0.0620	0.0250
		0.0274	0.0631	0.0256	0.0269	0.0626

(Continued)

Table 5.3 (Continued)

Design		Number of categories (c)				
(t, b, k, r)	bk	c = 3	c = 10	c = 20	c = 50	c = 100
(7, 21, 2, 3)	42	0.0004	0.0217	0.0587	0.0088	0.0242
		0.0608	0.0133	0.0218	0.0647	0.0157
		0.0194	0.0681	0.0166	0.0185	0.0684
(6, 15, 4, 10)	60	0.0078	0.0418	0.0529	0.0298	0.0425
		0.0536	0.0359	0.0424	0.0535	0.0392
		0.0418	0.0528	0.0402	0.0416	0.0528
(6, 20, 3, 10)	60	0.0052	0.0405	0.0532	0.0266	0.0412
		0.0536	0.0331	0.0412	0.0537	0.0371
		0.0410	0.0543	0.0383	0.0407	0.0552

The first entry in each cell uses the unadjusted Durbin test, the second the adjusted Durbin, and the third the ANOVA F test.

is never satisfactory for $c = 3$ and is barely so for $bk = 60$ and the larger c. The adjusted Durbin test is only satisfactory for the designs with $bk = 60$.

In the light of these results, we recommend inference be based on the ANOVA F test on the ranks when bk is greater than 20. Otherwise resampling methods should be used.

Our set-up reflects how data are obtained when judges (blocks) assign scores to treatments, as in a Likert study, and for each judge scores are ranked. We acknowledge that other set-ups are possible.

For an comparison of power for the competing tests see Livingston Jr. and Rayner (2022).

5.5 Orthogonal Contrasts for Balanced Designs with Ordered Treatments

5.5.1 Orthogonal Contrasts

Frequently used in the analysis of variance, a contrast is a linear combination of random variables whose coefficients add to zero. Contrasts allow a comparison between the variables in the contrast. So if $X_1, \ldots, X_t$ are t variables and if $c_1, \ldots, c_t$ are t constants that sum to zero then $c_1 X_1 + \cdots + c_t X_t$ is a contrast in those variables. If, for example, we are interested in how treatment one compares with the remainder of the treatments we could use the contrast $X_1 - \frac{(X_2 + \cdots + X_t)}{t-1}$.

In many important situations, we wish to test the null hypothesis of equal treatment means or equal treatment mean ranks. Suppose these means are $\mu_1, \ldots, \mu_t$ with $\mu = (\mu_1, \ldots, \mu_t)^{\mathrm{T}}$. If $\bar{\mu} = \frac{\sum_{i=1}^{t} \mu_i}{t}$, then we wish to test the null hypothesis $\mu = \bar{\mu} 1_t$ against various alternatives. If M is any $t \times t$ matrix of full rank, then $\mu = \bar{\mu} 1_t$ if and only if $M\mu = M\bar{\mu} 1_t$. Now if the last row of M is 1_t^{T}, then $1_t^{\mathrm{T}} \mu = 1_t^{\mathrm{T}} \bar{\mu} 1_t$ is just the identity defining the mean. If M is orthogonal with last row $\frac{1_t^{\mathrm{T}}}{\sqrt{t}}$ and $c_i^{\mathrm{T}} = (c_{i1}, \ldots, c_{it})$ is any other row, then by the orthogonality $c_{i1} + \cdots + c_{it} = 0$ and the $c_i^{\mathrm{T}} \mu$, for $i = 1, \ldots, t-1$ are orthogonal contrasts in the treatment means. A set of orthogonal contrasts that is often useful are the Helmert contrasts, based on the coefficients $\frac{1}{\sqrt{2}}, -\frac{1}{\sqrt{2}}, 0, \ldots, 0$ to compare the first two treatments, $\frac{2}{\sqrt{6}}, -\frac{1}{\sqrt{6}}, -\frac{1}{\sqrt{6}}, 0, \ldots, 0$ to compare the first treatment with the mean of the next two treatments, and so on. These sets of coefficients are normalised so that their sum of squares is 1.

Suppose M is orthogonal with rows $c_1^{\mathrm{T}}, \ldots, c_{t-1}^{\mathrm{T}}$ and $\frac{1_t^{\mathrm{T}}}{\sqrt{t}}$. The null hypothesis $\mu = \bar{\mu} 1_t$ is true if and only if $M\mu = M\bar{\mu} 1_t$ and hence if and only if all the 'component' null hypotheses $c_i^{\mathrm{T}} \mu = c_i^{\mathrm{T}} \bar{\mu} 1_t$, $i = 1, \ldots, t-1$ are true. For ordered treatments it is of interest to assess null hypotheses such as whether the treatments are non-decreasing (or non-increasing), or whether the treatments are non-decreasing then non-increasing (or, conversely, non-increasing then non-decreasing). Testing for such trend and umbrella effects can be achieved by choosing the rows of M to be orthonormal polynomials.

5.5.2 Orthogonal Contrasts for Nonparametric Testing in Balanced Designs

Thas et al. (2012) showed how to construct important orthogonal contrasts for the CRD, RBD, and the BIBD. However, in the case of the CRD, only the case where each treatment had the same number of observations - the balanced case - was considered. We will also focus on balanced designs. Our results also apply to the LSD, to be considered in Chapter 9, that is balanced.

It is well-known that if $Y = (Y_1, \ldots, Y_t)^{\mathrm{T}}$ is asymptotically multivariate normal with mean zero and covariance matrix Σ, then $Y^{\mathrm{T}} \Sigma^- Y$ is distributed as χ_r^2 in which Σ^- is a generalised inverse of Σ, such as the Moore–Penrose inverse, and r is the rank of Σ. Because the quadratic form $Y^{\mathrm{T}} \Sigma^- Y$ is distributed as χ_r^2 both Σ and Σ^- are idempotent with r eigenvalues one and $t - r$ eigenvalues zero. By the singular value decomposition theorem $\Sigma^- = c_1 c_1^{\mathrm{T}} + \cdots + c_r c_r^{\mathrm{T}}$ where $c_1, \ldots, c_r$ are the normalised eigenvectors corresponding to the non-zero eigenvalues. Thus $\Sigma^- c_i = c_i$, for $i = 1, \ldots, r$. It follows that $Y^{\mathrm{T}} \Sigma^- Y = (c_1^{\mathrm{T}} Y)^2 + \cdots + (c_r^{\mathrm{T}} Y)^2$.

The c_i are mutually orthonormal, and if we define the $t \times r$ matrix $C = (c_1 | \dots | c_r)$ then $C^T C = C^T \Sigma C = I_r$. It now follows that $\text{Cov}\left(c_i^T Y, c_j^T Y\right) = \left(C^T \Sigma C\right)_{ij} = \delta_{ij}$, and under the null hypothesis of no treatment effects, the $c_i^T Y$ are mutually independent and standard normal. They are also contrasts if $c_i^T 1_t = 0$. With this choice the $c_i^T Y$ are orthogonal contrasts with the important properties that their sum of squares is $Y^T \Sigma^- Y$ and the squares themselves are, again under the null hypothesis of no treatment effects, mutually independent χ_1^2 variables. The squared contrasts decompose the quadratic form $Y^T \Sigma^- Y$.

For the RBD and the BIBD if the data are appropriate ranks, the quadratic forms are the Friedman and Durbin test statistics, and $r = t - 1$. In these cases, Y is the vector of standardised treatment rank sums.

The null hypothesis is that the treatment mean ranks are equal, all being μ, say. There is a single linear constraint, $1_t^T Y = 0$. This suggests that $\text{E}\left[\left(1_t^T Y\right)\left(1_t^T Y\right)^T\right] = 1_t^T \Sigma 1_t = 0$ and zero is an eigenvalue with eigenvector 1_t. All other eigenvalues are 1 and the only constraints on their eigenvectors are that they are orthogonal to 1_t and normalised. If $1_t^T Y \neq 0$ then $\left(1_t^T Y\right)^2 > 0$ and $\text{E}\left[\left(1_t^T Y\right)\left(1_t^T Y\right)^T\right] = 1_t^T \Sigma 1_t > 0$ so that 1_t is not an eigenvector, and 1 is not an eigenvalue.

Orthogonal contrasts can be used to test important hypotheses. Under the null hypothesis that the mean treatment ranks are equal if, for example, $c_1^T = \left(1 - \frac{(t+1)}{2}, 2 - \frac{(t+1)}{2}, \dots, t - \frac{(t+1)}{2}\right)$, the contrast $c_1^T Y$ may be used to test whether the mean treatment ranks are non-decreasing (or non-increasing). For the RBD this is Page's test; for the BIBD it is called a Page-like test. There are other ways in which the null hypothesis may fail; for example, umbrella effects are sometimes of interest. An umbrella effect occurs when the rank means are either non-decreasing then non-increasing or, conversely, non-increasing then non-decreasing.

To give the $c_i^T Y$ polynomial interpretations like those just described, one choice when there are no ties is to base the c_i on the orthonormal discrete Legendre polynomials. In general for random variable X with mean μ and central moments μ_r, $r = 2, 3, \dots$, we use the following system. The polynomial of degree zero is taken to be identically one: $a_0(x) = 1$ for all x. The orthonormal polynomials of first and second degree are

$$a_1(x) = \frac{x - \mu}{\sqrt{\mu_2}} \text{ and}$$

$$a_2(x) = \frac{(x-\mu)^2 - \frac{\mu_3(x-\mu)}{\mu_2} - \mu_2}{\sqrt{d}} \text{ in which } d = \mu_4 - \frac{\mu_3^2}{\mu_2} - \mu_2^2.$$

More detail on orthogonal polynomials is given in Section 6.2. Now the initial moments of the discrete uniform distribution on $1, 2, \ldots, t$ are

$$\mu = \frac{t+1}{2}, \quad \mu_2 = \frac{t^2-1}{12}, \quad \mu_3 = 0, \quad \text{and}$$

$$\mu_4 = \frac{(t-1)(t+1)(3t^2-7)}{240}.$$

The vector c_i of contrast coefficients has elements $c_i(x)$. From the normality of the $\{a_i(x)\}$ we have $E\left[a_i^2(x)\right] = 1$ and need $c_i^T(x)c_i(x) = 1$, so we put $c_i(x) = \frac{a_i(x)}{\sqrt{t}}$. Then for $i = 1, 2$:

$$c_1(x) = \left(x - \frac{t+1}{2}\right)\sqrt{\frac{12}{t(t^2-1)}} \text{ and}$$

$$c_2(x) = \left[\left(x - \frac{t+1}{2}\right)^2 - \frac{t^2-1}{12}\right]\sqrt{\frac{180}{(t^2-1)(t^2-4)}}.$$

A table of the linear and quadratic coefficients for $t = 3, \ldots, 7$ is given in Table 5.4 in which $c_i(x) = c_{ix}$.

In most applications major interest is in linear and quadratic orthogonal contrasts, and rather than calculate further contrasts, those of third degree and higher are usually aggregated. A remainder $Y^T \Sigma^- Y - \left(c_1^T Y\right)^2 - \left(c_2^T Y\right)^2$ may be used to assess higher order effects. We call the $c_i^T Y$ χ^2 contrasts as opposed to F contrasts that will be introduced in Section 5.5.3.

There are other possible choices for the $\{c_i\}$. However $\{c_i^T \mu\}$ forms a basis for the parameter space and while any increasing sequence of constants may certainly be used in c_1 to construct a test statistic, all such test statistics will be sensitive to an increasing trend. Similarly, if it is known a priori that, say, the third of nine response categories is most likely to provide the minimum response, then the quadratic coefficients can be chosen to reflect that and potentially achieve more power.

For a treatment of orthogonal contrasts in unbalanced designs, see Rayner and Livingston Jr. (2022).

5.5.2.1 The RBD Example

Assume we have b blocks and t treatments, and suppose that r_{ij} is the rank given to treatment i on block j, that R_i is the sum of the ranks for treatment i and that $\sum_{i=1}^{t}\sum_{j=1}^{b} r_{ij}^2 - \frac{bt(t+1)^2}{4} = t\sigma^2$, say, is the sum of the rank variances over the b blocks. Then the Friedman test statistic adjusted for ties is

$$Fdm_A = \frac{(t-1)\left[\sum_{i=1}^{t}\left(R_i - \frac{b(t+1)}{2}\right)^2\right]}{t\sigma^2}.$$

Table 5.4 Linear and quadratic coefficients.

(a) Linear coefficients

t	$c_{11}, c_{12}, \ldots, c_{1t}$
3	$-\frac{1}{\sqrt{2}}, 0, \frac{1}{\sqrt{2}}$
4	$-\frac{3}{\sqrt{20}}, -\frac{1}{\sqrt{20}}, \frac{1}{\sqrt{20}}, \frac{3}{\sqrt{20}}$
5	$-\frac{2}{\sqrt{10}}, -\frac{1}{\sqrt{10}}, 0, \frac{1}{\sqrt{10}}, \frac{2}{\sqrt{10}}$
6	$-\frac{5}{\sqrt{70}}, -\frac{3}{\sqrt{70}}, -\frac{1}{\sqrt{70}}, \frac{1}{\sqrt{70}}, \frac{3}{\sqrt{70}}, \frac{5}{\sqrt{70}}$
7	$-\frac{3}{\sqrt{28}}, -\frac{2}{\sqrt{28}}, -\frac{1}{\sqrt{28}}, 0, \frac{1}{\sqrt{28}}, \frac{2}{\sqrt{28}}, \frac{3}{\sqrt{28}}$

(b) Quadratic coefficients

t	$c_{21}, c_{22}, \ldots, c_{2t}$
3	$\frac{1}{\sqrt{6}}, -\frac{2}{\sqrt{6}}, \frac{1}{\sqrt{6}}$
4	$\frac{1}{\sqrt{4}}, -\frac{1}{\sqrt{4}}, -\frac{1}{\sqrt{4}}, \frac{1}{\sqrt{4}}$
5	$\frac{2}{\sqrt{14}}, -\frac{1}{\sqrt{14}}, -\frac{2}{\sqrt{14}}, -\frac{1}{\sqrt{14}}, \frac{2}{\sqrt{14}}$
6	$\frac{5}{\sqrt{84}}, -\frac{1}{\sqrt{84}}, -\frac{4}{\sqrt{84}}, -\frac{1}{\sqrt{84}}, \frac{5}{\sqrt{84}}$
7	$\frac{5}{\sqrt{84}}, 0, -\frac{3}{\sqrt{84}}, -\frac{4}{\sqrt{84}}, -\frac{3}{\sqrt{84}}, 0, \frac{5}{\sqrt{84}}$

Now if

$$Y_i = \left[R_i - \frac{b(t+1)}{2} \right] \sqrt{\frac{t-1}{t\sigma^2}},$$

then $Fdm_A = \sum_{i=1}^{t} Y_i^2$. Since $1_t^T Y = 0$ the $c_i^T Y$ are the χ^2 orthogonal contrasts for the RBD. In particular $c_1^T Y$ is the Page test statistic. If there are no ties $Y_i = \left[R_i - \frac{b(t+1)}{2} \right] \sqrt{\frac{12}{[bt(t+1)]}}$.

5.5.2.2 The BIBD Example
Durbin's statistic adjusted for ties is

$$D_A = c \sum_{i=1}^{t} \left[R_i - \frac{r(k+1)}{2} \right]^2,$$

in which

$$c = \frac{t-1}{-\frac{bk(k+1)^2}{4} + \sum_{i=1}^{t} \sum_{j=1}^{b} r_{ij}^2}.$$

So if

$$Y_i = \left[R_i - \frac{r(k+1)}{2} \right] \sqrt{c},$$

then $D_A = \sum_{i=1}^{t} Y_i^2$. Now $1_t^T Y = 0$ so the $c_i^T Y$ are the orthogonal contrasts. If there are no ties then $c = \frac{12(t-1)}{rt(k^2-1)}$.

5.5.3 *F* Orthogonal Contrasts

Although the treatment following applies to any data, we are mainly interested in analysing ranks. In many fixed effects general linear models, the treatment sum of squares SS_{Treat} is proportional to

$$\sum_{i=1}^{t} \frac{R_i^2}{n_i} - \frac{T^2}{n} = \sum_{i=1}^{t} Z_i^2,$$

in which $Z_i = \frac{R_i}{\sqrt{n_i}} - T\frac{\sqrt{n_i}}{n} = \frac{R_i - T\frac{n_i}{n}}{\sqrt{n_i}}$ for $i = 1, \ldots, t$, with R_i the sum of the ranks for the ith of the t treatments being ranked and $T = \sum_{i=1}^{t} R_i$, the grand total of the ranks. For the time being the proportionality constant will be ignored. Designs for which all n_i are equal are called balanced designs. The χ^2 orthogonal contrasts of Section 5.5.2 and the F orthogonal contrasts constructed here require the design to be balanced.

Note that in balanced designs $\sum_{i=1}^{t} Z_i = 0$ so if $Z = (Z_1, \ldots, Z_t)^T$ then $1_t^T Z = 0$. If σ^2 is the expectation of the error mean square MS_{Error} then SS_{Treat} is distributed as $\sigma^2 \chi_{t-1}^2$. As is usual we test for equality of treatment rank means using $F = \frac{\frac{SS_{\text{Treat}}}{t-1}}{MS_{\text{Error}}}$.

Now as SS_{Treat} is distributed as $\sigma^2 \chi_{t-1}^2$ then as in Section 5.5.2 there exist orthogonal columns m_s, $s = 1, \ldots, t-1$, such that $SS_{\text{Treat}} = \left(m_1^T Z \right)^2 + \cdots + \left(m_{t-1}^T Z \right)^2$ and the $\left(m_s^T Z \right)^2$ are mutually independent $\sigma^2 \chi_1^2$ variables. The $\frac{\left(m_s^T Z \right)^2}{MS_{\text{Error}}}$ may be referred to the $F_{1, \text{edf}}$ distribution to test if each $E \left[m_s^T Z \right]$ is significantly different from zero. We call these $\frac{\left(m_s^T Z \right)^2}{MS_{\text{Error}}}$ the F contrasts.

If the original null hypothesis is that rank means are equal and the orthogonal columns are based on orthonormal polynomials, then these contrasts assess linear, quadratic etc. differences between the mean ranks. Usually effects after the second are aggregated into a remainder $SS_{\text{Treat}} - \left(m_1^T Z \right)^2 - \left(m_2^T Z \right)^2$, which, when divided by the degrees of freedom, $t - 3$, gives a remainder mean square. Subsequently dividing by MS_{Error} gives a remainder statistic that has the $F_{t-3, \text{edf}}$ distribution. For a small number of treatments, some of the contrasts may not be characterised as polynomial effects. Since the rth polynomial is orthogonal to those of degree $0, 1, \ldots, r - 1$, there are $1 + 2 + \cdots + r = \frac{r(r+1)}{2}$ constraints on the m_s. To have a quadratic effect $(r = 2)$ requires $\frac{r(r+1)}{2} = 3$ constraints and at least three treatments. To have a cubic effect $(r = 3)$ requires $\frac{r(r+1)}{2} = 6$ constraints and at least six treatments. Thus for three treatments we can have linear and quadratic effects; for four and five treatments we can have linear and quadratic effects and a remainder that can't be characterised as a polynomial effect or an aggregation of polynomial effects. These considerations apply equally to the χ^2 orthogonal contrasts.

5.5.3.1 The RBD Example
For this design $Z_i = \dfrac{R_i - \frac{b(t+1)}{2}}{\sqrt{b}}$ for $i = 1, \ldots, t$.

5.5.3.2 Lemonade Taste Example
Thas et al. (2012, Section 4.2) give the data in Table 5.5. They decompose the adjusted Friedman statistic adjusted for ties into Page, umbrella, and remainder components with χ^2 p-values 0.68, 0.03, and 0.22. Our calculations give χ_1^2 p-values 0.6831, 0.0285, and 0.2207, respectively. The unadjusted Friedman p-value is given as 0.09.

Routine calculation finds the treatments sum of squares, SS_{Treat}, to be 9.7 on 3 degrees of freedom and the error mean square, MS_{Error}, is 1.0667 on 12 degrees of freedom. The treatments p-value is 0.0710.

We now decompose SS_{Treat}, into linear, quadratic, and remainder sums of squares and test each using the F distribution with 1 and edf degrees of freedom. First, the ranks sums for treatments A–D are 14, 11.5, 7.5, and 17, respectively. Hence $Z = \frac{(1.5,-1,-5,4.5)}{\sqrt{5}}$ from which $SS_{\text{Treat}} = Z^{\text{T}} Z = 9.7$. The F contrasts are 0.2344, 6.75, and, by difference, 2.1094. The corresponding F p-values are 0.6370, 0.0233, and 0.1720.

Table 5.5 Lemonades ranked by five tasters.

| | Lemonades | | | | Rank |
Taster	A	B	C	D	sum
1	3	2	1	4	10
2	3	1.5	1.5	4	10
3	1	4	2	3	10
4	3	2	1	4	10
5	4	2	2	2	10
Rank sum	14	11.5	7.5	17	50

Although the Durbin test and corresponding ANOVA F test on the ranks are not significant at the 0.05 level, the quadratic orthogonal contrasts for both tests are. The more focused contrasts are able to detect an effect the more omnibus tests cannot.

```
friedman(y = rank, groups = type, blocks = taster, components = T)

##
##           Friedman Rank Sum Test
##
## Test statistic is adjusted for ties
## data:   rank type taster
## Fdm_A = 6.466667, df = 3, p-value = 0.0909864
##
## For the group levels A B C D,
## the rank sums are:   14 11.5 7.5 17
##
##
## Chi-square components p-values
## Overall:   0.09099
## Linear:    0.68309
## Quadratic:   0.02846
## Remainder:   0.22067
##
## F components p-values
## Overall:   0.071
## Linear:    0.63701
## Quadratic:   0.02331
## Remainder:   0.17204
```

5.5.3.3   The BIBD Example

$$Z_i = \left[R_i - \frac{r(k+1)}{2} \right] \sqrt{\frac{k}{\lambda t}}.$$

5.5.3.4   Ice Cream Flavouring Data

Consider again the data in Table 5.1 relating to seven vanilla ice creams. We previously found that Durbin's D has χ^2 p-value 0.0620, but the equivalent ANOVA on the ranks has p-value 0.0049. The simulation study in Section 5.4 demonstrated that the latter is uniformly superior to the former. This suggests that the F contrasts will likewise be superior to the χ^2 contrasts. A brief simulation study in Section 5.5.4 considers this.

Since we have no ties $c = \frac{12(t-1)}{rt(k^2-1)} = \frac{3}{7}$. The rank sums are 8, 9, 4, 3, 5, 6, and 7 so $Y = \left[R_i - \frac{r(k+1)}{2} \right] \sqrt{\frac{k}{\lambda t}} = (2,3,-2,-3,-1,0,1) \sqrt{\frac{3}{7}}$. In this case $Y = Z$ and $Z^T Z = SS_{\text{Treat}} = 12.0 = D$. The contrasts are exactly as for χ^2 analysis but now their p-values are found using the F distribution.

Given the unreliability of the χ^2 p-values it is of interest to also calculate permutation test p-values. Those given here are based on 1,000,000 permutations. We find a Durbin permutation test p-value of 0.0000. The Durbin component p-values based on the permutation test are 0.3066,

0.0056, and 0.3790 for the linear, quadratic, and remainder components, respectively. The overall F permutation p-value is 0.0007 with component p-values of 0.0857, 0.0011, and 0.048, respectively. For completeness the p-values of the χ^2 components are 0.3223, 0.0101, and 0.3536 compared with the F contrasts p-values of 0.0831, 0.0009, and 0.0356.

The analysis by decomposing Durbin's statistic using χ^2 p-values identifies the quadratic effect as the chief explanation of the differences in rank means. The alternative analyses view the effect as far more complex.

```
durbin(y = rank, group = variety, blocks = judge, components = T)

##
##         Durbin Rank Sum Test
##
## Test statistic is adjusted for ties
## data:   rank variety judge
## D_A = 12, df = 6, p-value = 0.0619688
##
## For the group levels 1 2 3 4 5 6 7,
## the rank sums are:  8 9 4 3 5 6 7
##
##
## Chi-square components p-values
## Overall:   0.06197
## Linear:    0.3223
## Quadratic:  0.01013
## Remainder:  0.35358
##
## F components p-values
## Overall:   0.0049
## Linear:    0.08311
## Quadratic:  0.00088
## Remainder:  0.0356
```

5.5.4 Simulation Study

In the following simulation study we calculate test sizes as the proportion of rejections using the linear, quadratic, and remainder components of the Durbin test together with the same components for the ANOVA F test on the ranks. We do so at the 5% significance level and use the same set of designs as outlined in Section 5.4.

The results from 1,000,000 simulations are shown in Table 5.6. It appears as though the test sizes for the χ^2 components tend to be below the nominal 5% level where as the F component sizes are in the previous text. For smaller sample sizes there is greater deviation from the nominal significance level.

Table 5.6 Proportion of rejections using chi-squared and F contrasts for a nominal significance level of 0.05 based on 1,000,000 simulations.

Design (t, b, k, r)	bk	$c = 3$	$c = 10$	$c = 20$	$c = 50$	$c = 100$
			Number of categories (c)			
(4, 6, 2, 3)	12	0.0221	0.0306	0.031	0.0312	0.0315
		0.0364	0.0833	0.1021	0.1144	0.1196
		0.0226	0.0303	0.0312	0.0311	0.0312
		0.0700	0.0581	0.0469	0.0383	0.0352
		0.0812	0.1054	0.1141	0.1193	0.1222
		0.0703	0.0580	0.0471	0.0381	0.0348
(4, 4, 3, 3)	12	0.0462	0.0444	0.0413	0.0379	0.0363
		0.0514	0.0429	0.0340	0.0282	0.0267
		0.0461	0.0445	0.0415	0.0380	0.0362
		0.0628	0.0738	0.0785	0.0828	0.0844
		0.0572	0.0606	0.0557	0.0497	0.0471
		0.0626	0.0734	0.0785	0.0826	0.0844
(5, 10, 2, 4)	20	0.0473	0.0483	0.0537	0.0592	0.0613
		0.0484	0.0433	0.0388	0.0349	0.0335
		0.0310	0.0349	0.0359	0.0374	0.0379
		0.0588	0.0643	0.0681	0.0713	0.0726
		0.0594	0.0638	0.0642	0.0640	0.0643
		0.0620	0.0705	0.0729	0.0753	0.0761
(5, 5, 4, 4)	20	0.0465	0.0468	0.0479	0.0482	0.0482
		0.0485	0.0489	0.0483	0.0466	0.0463
		0.0407	0.0415	0.0413	0.0413	0.0412
		0.0541	0.0548	0.0554	0.0557	0.0560
		0.0540	0.0550	0.0549	0.0546	0.0549
		0.0572	0.0589	0.0587	0.0586	0.0583
(7, 7, 3, 3)	21	0.0476	0.0475	0.0477	0.0507	0.0526
		0.0492	0.0490	0.0490	0.0479	0.0474
		0.0436	0.0440	0.0433	0.0425	0.0421
		0.0524	0.0524	0.0521	0.0524	0.0524
		0.0518	0.0523	0.0516	0.0504	0.0495
		0.0545	0.0549	0.0549	0.0554	0.0555

(Continued)

Table 5.6 (Continued)

Design (t, b, k, r)	bk	Number of categories (c)				
		$c = 3$	$c = 10$	$c = 20$	$c = 50$	$c = 100$
(7, 7, 4, 4)	28	0.0473	0.0484	0.0484	0.0481	0.048
		0.0478	0.0445	0.0434	0.0423	0.0424
		0.0312	0.0325	0.0325	0.0319	0.0319
		0.0526	0.0534	0.0531	0.0510	0.0501
		0.0534	0.0541	0.0537	0.0526	0.0529
		0.0570	0.0601	0.0609	0.0607	0.0613
(5, 10, 3, 6)	30	0.0492	0.0483	0.0489	0.0478	0.0473
		0.0475	0.0474	0.0499	0.0539	0.0559
		0.0378	0.0387	0.0390	0.0389	0.0391
		0.0520	0.0522	0.0533	0.0537	0.0541
		0.0525	0.0525	0.0527	0.0524	0.0525
		0.0556	0.0568	0.0569	0.0569	0.0568
(6, 15, 2, 5)	30	0.0494	0.0492	0.0488	0.0493	0.0494
		0.0497	0.0495	0.0497	0.0478	0.0476
		0.0456	0.0456	0.0454	0.0458	0.0453
		0.0508	0.0506	0.0503	0.0509	0.0509
		0.0508	0.0506	0.0508	0.0502	0.0511
		0.0520	0.0519	0.0519	0.0520	0.0517
(6, 10, 3, 5)	30	0.0487	0.0487	0.0488	0.0478	0.0472
		0.0487	0.0486	0.0489	0.0509	0.0531
		0.0448	0.0454	0.0454	0.0450	0.0453
		0.0504	0.0503	0.0504	0.0504	0.0509
		0.0506	0.0505	0.0511	0.0507	0.0509
		0.0521	0.0529	0.0526	0.0521	0.0525
(7, 21, 2, 3)	42	0.0465	0.0477	0.0476	0.0492	0.0509
		0.0471	0.0485	0.0483	0.0475	0.0468
		0.0191	0.0202	0.0203	0.0203	0.0205
		0.0553	0.0563	0.0561	0.0551	0.0557
		0.0552	0.0556	0.0562	0.0563	0.0561
		0.0612	0.0640	0.0646	0.0651	0.0658

Table 5.6   (Continued)

Design		Number of categories (c)				
(t, b, k, r)	bk	c = 3	c = 10	c = 20	c = 50	c = 100
(6, 15, 4, 10)	60	0.0481	0.0479	0.0475	0.047	0.0456
		0.0482	0.0484	0.0479	0.0477	0.0481
		0.0332	0.0339	0.0339	0.0337	0.0338
		0.0523	0.0527	0.0526	0.0527	0.0524
		0.0523	0.0524	0.0525	0.0526	0.0530
		0.0566	0.0573	0.0574	0.0573	0.0572
(6, 20, 3, 10)	60	0.0489	0.0488	0.0496	0.0514	0.0517
		0.0484	0.0478	0.0487	0.0505	0.0506
		0.0326	0.0339	0.0340	0.0336	0.0339
		0.0521	0.0527	0.0520	0.0526	0.0524
		0.0520	0.0524	0.0524	0.0525	0.0517
		0.0556	0.0571	0.0571	0.0568	0.0573

The first three entries in each cell are for the linear, quadratic, and remainder components of the adjusted Durbin statistic. The following three entries are for linear, quadratic, and remainder components of the ANOVA F statistic. These are results for values $bk \leq 28$ and $bk \geq 30$.

While the overall ANOVA F test on the ranks clearly outperformed the overall Durbin test, the difference between the component tests here are not so obvious. It does seem the χ^2 remainder tests deviate from the nominal significance level slightly more than the remainder tests for the ANOVA F. With this in mind we would recommend the use of the F component tests using permutation tests.

5.6   A CMH MS Analogue Test Statistic for the BIBD

Recall that in part, the objective in this chapter was to see how CMH methods can aid analysis for incomplete designs. For the BIBD the CMH mean scores (MS) statistic cannot be directly calculated: the BIBD is not a CMH design. However when there are no ties the Durbin test can be shown to be an analogue of the CMH MS test.

It was shown in Section 2.4 that for any design consistent with the CMH methodology the CMH MS test statistic T_{MS} can be calculated as follows.

First put $V = \left(M_{i\bullet} - \mathrm{E}\left[M_{i\bullet}\right]\right)$. Then $T_{\mathrm{MS}} = V^{\mathrm{T}}\left[\sum_{j=1}^{b} \mathrm{Cov}\left(M_j\right)\right]^{-} V$ in which, if $p_{i\bullet j} = \frac{n_{i\bullet j}}{n_{\bullet\bullet j}}$, then $\mathrm{Cov}\left(M_j\right) = S_j^2\left(\mathrm{diag}\left(p_{i\bullet j}\right) - \left\{p_{i'\bullet j}p_{i\bullet j}\right\}\right)$.

For the BIBD assume we have ranks with no ties. Then on each block the scores are $1, 2, \ldots, k$: $b_{hj} = h$ for $1, \ldots, k$ for all j. On the jth block there is at most one observation of treatment i, so $n_{ihj} = 1$ at most once, and is otherwise zero. Moreover $n_{\bullet\bullet j} = k$, and $(k-1)S_j^2 = k\left(1^2 + \cdots + k^2\right) - (1 + \cdots + k)^2 = \frac{k^2(k^2-1)}{12}$, so that $S_j^2 = \frac{k^2(k+1)}{12}$ for all j. Write $S_j^2 = S^2$ for all j.

Now $M_{i\bullet}$ is the score sum over blocks for treatment i, which here is R_i, the rank sum for treatment i. This has expectation $\mathrm{E}\left[R_i\right] = \frac{r(k+1)}{2}$, since, for any given i, there are r blocks containing that treatment and each block total is $\frac{k(k+1)}{2}$. Hence the treatment average for all treatments on that block is $\frac{(k+1)}{2}$ and the average across blocks that contain treatment i is $\frac{r(k+1)}{2}$.

The rank sum on each of b blocks is $\frac{k(k+1)}{2}$, and hence $\sum_{i=1}^{t} R_i = \frac{bk(k+1)}{2}$ overall. We have just shown in the preceding paragraph that $\mathrm{E}\left[R_i\right] = \frac{r(k+1)}{2}$, so $\sum_{i=1}^{t} \mathrm{E}\left[R_i\right] = \frac{bk(k+1)}{2}$. Thus $1_t V = \sum_{i=1}^{t}\left(R_i - \mathrm{E}\left[R_i\right]\right) = 0$.

On the jth block temporarily ignore the fact that treatments that are missing. If the treatments present are numbered $1, \ldots, k$ then $p_{i\bullet j} = \frac{n_{i\bullet j}}{n_{\bullet\bullet j}} = \frac{1}{k}$, and $\mathrm{diag}\left(p_{i\bullet j}\right) - \left\{p_{i'\bullet j}p_{i\bullet j}\right\} = \frac{\left\{I_k - \frac{1_k 1_k^{\mathrm{T}}}{k}\right\}}{k}$. Thus $\mathrm{Cov}\left(M_j\right) = S^2\frac{\left\{I_k - \frac{1_k 1_k^{\mathrm{T}}}{k}\right\}}{k}$.

Although the treatments differ from block to block, the elements of $\mathrm{Cov}\left(M_j\right)$ corresponding to the ith treatment are identical on the r blocks containing that treatment. Thus

$$\mathrm{Cov}(M) = \sum_{j=1}^{b}\mathrm{Cov}\left(M_j\right) = rS^2\frac{\left\{I_k - \frac{1_k 1_k^{\mathrm{T}}}{k}\right\}}{k} \quad \text{and}$$

$$\mathrm{Cov}^{-}(M) = \frac{k\left\{I_k - \frac{1_k 1_k^{\mathrm{T}}}{k}\right\}}{rS^2}.$$

Accumulating these results, since $\frac{S^2}{k} = \frac{k(k+1)}{12}$,

$$T_{\mathrm{MS}} = V^{\mathrm{T}}\left[\sum_{j=1}^{b}\mathrm{Cov}\left(M_j\right)\right]^{-} V = k\sum_{i=1}^{t}\frac{V_i^2}{rS^2} = 12\sum_{i=1}^{t}\frac{V_i^2}{rk(k+1)}.$$

It is perhaps better to denote this statistic by T_{AMS}. It is not the CMH MS statistic, but an analogue of it. Neither is it the Durbin statistic; in fact $\frac{T_{\mathrm{AMS}}}{D} = \frac{b(k-1)}{r(t-1)}$. For BIBDs that occur in practice it seems that frequently, although not always, $T_{\mathrm{AMS}} < D$, so the test based on T_{AMS} is likely to be even more conservative than that based on D.

While we cannot conclude that the Durbin test is a CMH MS test, we can conclude that when there are no ties it is an analogue of such a test. The significance of this result is that we are only able to claim the ANOVA F test is a competitor for the Durbin test, not an alternative form of it.

It would seem that CMH methods have not led to further insights in the BIBD. However we will find subsequently that the nonparametric ANOVA approach does! In particular see Chapters 8 and 10

5.6.1 Ice Cream Flavouring Data

For these data $b = t = 7$, $k = r = 3$. Thus $\frac{b(k-1)}{r(t-1)} = 7 \times \frac{2}{3 \times 6}$ and $T_{AMS} = \frac{7D}{9} = \frac{28}{3}$ with χ_6^2 p-value 0.1557.

Bibliography

Best, D. J. and Rayner, J. C. W. (2014). Conover's F test as an alternative to Durbin's test. *Journal of Modern Applied Statistical Methods*, 13(2):4.

Brockhoff, P. B., Best, D. J., and Rayner, J. C. W. (2004). Partitioning Anderson's statistic for tied data. *Journal of Statistical Planning and Inference*, 121(1):93–111.

Conover, W. J. (1998). *Practical Nonparametric Statistics*. New York: John Wiley & Sons.

Kuehl, R. (2000). *Design of Experiments: Statistical Principles of Research Design and Analysis*. Belmont, CA: Duxbury Press.

Kutner, M. H., Nachtsheim, C. J., Neter, J., and Li, W. (2005). *Applied Linear Statistical Models*. New York: McGraw-Hill.

Livingston, G. C. and Rayner, J. C. W. (2022). Nonparametric analysis of balanced incomplete block rank data. *J Stat Theory Pract.* 16:66.

Rayner, J. C. W. (2016). *Introductory Nonparametrics*. Copenhagen: Bookboon.

Rayner, J. C. W. and Livingston Jr., G. C. (2022). Orthogonal contrasts for both balanced and unbalanced designs and both ordered and unordered treatments. NIASRA Statistics Working Paper Series, 05-22.

Thas, O., Best, D. J., and Rayner, J. C. W. (2012). Using orthogonal trend contrasts for testing ranked data with ordered alternatives. *Statistica Neerlandica*, 66(4):452–471.

6

Unconditional Analogues of CMH Tests

6.1 Introduction

In Chapter 2 we defined and described the basic CMH tests: for overall partial association (OPA), general association (GA), mean scores (MS), and correlation (C). It was observed that these tests were *conditional* tests, derived under the assumptions that in each stratum the row and column totals are known. Thus the treatment totals within strata, $n_{i\bullet j}$, and the outcome category totals within strata, $n_{\bullet hj}$, and hence also the strata totals, $n_{\bullet\bullet j}$, are all assumed to be known prior to sighting the data: they are not random variables. Therefore under the null hypothesis of no association between the responses and the treatments, the counts N_{ihj} can be modelled by the *product extended hypergeometric distribution* that assumes fixed marginal totals.

The unconditional tests developed in this chapter assume that while the $n_{i\bullet j}$ and $n_{\bullet\bullet j}$ are known before sighting the data, the outcome category totals within strata are not known, and should therefore be written as $N_{\bullet hj}$ rather than $n_{\bullet hj}$. The convention of writing capitals for random variables and lower case for particular values of random variables is sometimes overlooked in literature.

For many data sets of interest, even though the outcome totals within strata are not known, the CMH methodology is nevertheless applied. Inference is then *conditional* upon those totals taking the observed values. *Conditional inference* is a valid statistical paradigm, and, like using permutation tests and the bootstrap, will often give inference that is very similar to *unconditional inference*. This is indeed the case in many of the examples we have looked at, as shall be seen later in this chapter. The agreement will be comforting to those statisticians more familiar with unconditional inference. A lack of agreement would certainly be disappointing – that different conclusions can be reached by using different paradigms of inference. However, this is certainly possible: see the cross-over clinical data

An Introduction to Cochran–Mantel–Haenszel Testing and Nonparametric ANOVA, First Edition. J.C.W. Rayner and G. C. Livingston Jr.
© 2023 John Wiley & Sons Ltd. Published 2023 by John Wiley & Sons Ltd.

example in Section 6.5. The implication is that inference may depend on the paradigm of inference and possibly also on other choices. It is thus important to make such choices carefully and frame the conclusions appropriately.

In Chapter 2 we distinguished between the nominal (OPA and GA tests) and ordinal (MS and C tests) CMH tests. Alternatives for the nominal tests were suggested. It was noted that the CMH OPA test statistic S_{OPA} is the sum over strata of weighted Pearson X^2 statistics, the weights being $\frac{n_{\bullet\bullet j}-1}{n_{\bullet\bullet j}}$. This suggested using the unweighted sum of Pearson X^2 statistics, which was denoted by T_{OPA}.

There is a caveat we must raise to this suggestion. It is possible that, in some strata, marginal totals are not all positive, and hence the Pearson statistic for the jth strata, $X^2_{\text{P}j}$, say, is only defined if the categories with zero counts are dropped. Both S_{OPA} and T_{OPA} suffer from this problem.

6.1.1 Jams Data Revisited

These data were considered in Chapter 2 with the data given in Table 2.2. Three plum jams, A, B, and C, are given JAR sweetness codes by eight judges, ranging from 1 for not sweet enough, to 5 for too sweet.

This is a randomised block design with the judges as blocks. In calculating the Pearson statistic on any block, any category with zero marginal counts would usually be dropped. For example, the first judge gives scores of 2, 3, and 3. The first, fourth, and fifth response categories have zero counts: $n_{\bullet h1} = 0$ for $h = 1, 4$, and 5. Unless these categories are omitted X^2_{P1} is not defined. It is certainly possible to sum the $X^2_{\text{P}j}$ after making such an adjustment, but it is not clear how to interpret this statistic.

It was also noted in Chapter 2 that the CMH GA test statistic S_{GA} is a quadratic form with vector the treatment/outcome counts obtained by summing the N_{ihj} over strata. The covariance matrix of this vector is somewhat complicated to derive and hence is not often available in convenient software. It seems reasonable to base a test on the Pearson test for the table of counts $\{N_{ih\bullet}\}$. This statistic was denoted by T_{GA}. The tests based on T_{OPA} and T_{GA} are unconditional tests and both have the same asymptotic distributions as the corresponding conditional tests. Most users will have more familiarity with the unconditional tests and most packages will have routines for calculating Pearson tests even if they don't have routines for CMH tests.

The problem with marginal categories with zero counts, raised earlier, doesn't occur with the tests based on S_{GA} and T_{GA}.

We now turn our attention to the ordinal CMH tests. In Chapters 3 and 4 it was found that the familiar Kruskal–Wallis and Friedman tests were CMH

MS tests, although it was hardly evident that as such they were conditional tests. The same issue will resurface when we consider the Stuart-Maxwell tests later in this chapter. It is of interest to construct unconditional analogues of the CMH MS and C tests to see, for example, if they provide tests that are more powerful or more convenient than the corresponding CMH tests.

6.2 Unconditional Univariate Moment Tests

In the unconditional model the treatment totals are specified by the design and are therefore known before sighting the data, while the column totals, the number of responses in each response category, are not known. A reasonable model is that for $i = 1, \ldots, t$ the counts N_{ihj} on the jth stratum, $j = 1, \ldots, b$, follow a multinomial distribution: $(N_{i1j}, \ldots, N_{icj})$ is multinomial with total count $\sum_{h=1}^{c} N_{ijh} = n_{i \bullet h}$ and cell probabilities $(p_{i1j}, \ldots, p_{icj})$. For $h = 1, \ldots, c$, the maximum likelihood estimator of p_{ihj} is $\hat{p}_{ihj} = \frac{N_{ihj}}{n_{i \bullet j}}$.

For singly ordered two-way tables, an unconditional model is explored in Rayner and Best (2001, Chapter 4). To describe this model we need to first define orthonormal polynomials. The functions in $\{a_u(X)\}$ are *orthogonal* with respect to the distribution of X means that

$$E[a_u(X)a_v(X)] = 0,$$

for $u \neq v$.

If, in addition, $E[a_u^2(X)] = 1$ for all u then the functions have been normalised, and $\{a_u(X)\}$ is *orthonormal*. If, moreover, the functions are polynomials, then $\{a_u(X)\}$ is the set of orthonormal polynomials.

To explicitly give the orthonormal polynomials of a random variable X up to degree three, first write μ for the mean of X and $\mu_r, r = 2, 3, \ldots$ for the central moments of X. To avoid ambiguity we always take the polynomial of degree zero to be identically one: $a_0(x) = 1$ for all x. Then

$$a_1(x) = \frac{x - \mu}{\sqrt{\mu_2}}$$

$$a_2(x) = \frac{\left[(x - \mu)^2 - \frac{\mu_3(x - \mu)}{\mu_2} - \mu_2\right]}{\sqrt{d}}$$

$$a_3(x) = \frac{(x - \mu)^3 - a(x - \mu)^2 - b(x - \mu) - c}{\sqrt{e}},$$

(6.1)

where

$$d = \mu_4 - \frac{\mu_3^2}{\mu_2} - \mu_2^2,$$

$$a = \frac{\mu_5 - \frac{\mu_3 \mu_4}{\mu_2} - \mu_2 \mu_3}{d},$$

$$b = \frac{\frac{\mu_4^2}{\mu_2} - \mu_2 \mu_4 - \frac{\mu_3 \mu_5}{\mu_2} + \mu_3^2}{d},$$

$$c = \frac{2\mu_3 \mu_4 - \frac{\mu_3^3}{\mu_2} - \mu_2 \mu_5}{d}, \text{ and}$$

$$e = \mu_6 - 2a\mu_5 + \left(a^2 - 2b\right)\mu_4 + 2\left(ab - c\right)\mu_3 + \left(b^2 + 2ac\right)\mu_2 + c^2.$$

Polynomials of degree higher than three may be calculated using the recurrence relation of Emerson (1968). See also Rayner et al. (2008).

In practice, the $a_u(x)$ in (6.1) are functions with a scalar input x_i; however, the complete vector x is required to determine the central moments μ_r.

To give the results in Rayner and Best (2001, Chapter 4) in the context here, first define, on the jth stratum, orthonormal polynomials $\{a_{uj}(h)\}$ with respect to $(\hat{p}_{\bullet 1j}, \ldots, \hat{p}_{\bullet cj})$:

$$\sum_{h=1}^{c} a_{uj}(h) a_{vj}(h) \hat{p}_{\bullet hj} = \delta_{uv}.$$

As previously, we take $a_{0j}(h) = 1$ for all h. Subsequently the first and second degree orthonormal polynomials on the jth stratum, $a_{1j}(h)$ and $a_{2j}(h)$, will be particularly important. Next define

$$V_{uij} = \sum_{h=1}^{c} \frac{N_{ihj} a_{uj}(h)}{\sqrt{n_{i \bullet j}}}.$$

Rayner and Best (2001, Chapter 4), translated to the current context, show that

$$\sum_{u=1}^{c-1} \sum_{i=1}^{t} V_{uij}^2 = X_{Pj}^2,$$

the Pearson statistic on the jth stratum. There are further results there: the V_{uij} are score test statistics in their own right as are the $V_{u1j}^2 + \cdots + V_{utj}^2$. The latter gives a degree u assessment of the consistency of treatments in the jth stratum. It has asymptotic distribution χ_{t-1}^2. Summing these terms over strata gives $\sum_{i=1}^{t} \sum_{j=1}^{b} V_{uij}^2$, which, because of the mutual independence of the terms, has asymptotic distribution $\chi_{b(t-1)}^2$. This statistic gives a degree u assessment of the consistency of treatments over all b strata. In particular, $\sum_{i=1}^{t} \sum_{j=1}^{b} V_{1ij}^2$ gives an overall mean assessment of the consistency of treatments, $\sum_{i=1}^{t} \sum_{j=1}^{b} V_{2ij}^2$ gives an overall dispersion assessment of the consistency of treatments, and so on. The $\sum_{i=1}^{t} \sum_{j=1}^{b} V_{uij}^2$ partition $\sum_{j=1}^{b} X_{Pj}^2 = T_{OPA}$.

Average assessments may be obtained by summing over strata. It is incidentally shown in the proof of Rayner and Best (2001, Theorem 4.1) that the $\left(V_{u1j}, \ldots, V_{utj}\right)^{T}$ are asymptotically $(t-1)$-variate normal with mean zero and covariance matrix $I_t - \frac{f_j f_j^T}{n_{\bullet\bullet j}}$ in which $f_j = \left(\sqrt{n_{1\bullet j}}, \ldots, \sqrt{n_{t\bullet j}}\right)^{T}$.

Using this result the $\left(V_{u1\bullet}, \ldots, V_{ut\bullet}\right)^{T}$ are asymptotically $(t-1)$-variate normal with mean zero and covariance matrix $\sum_{j=1}^{b} \left(I_t - \frac{f_j f_j^T}{n_{\bullet\bullet j}}\right) = \Sigma$, say.

An average degree u assessment of the consistency of treatments can be based on the quadratic form

$$\left(V_{u1\bullet}, \ldots, V_{ut\bullet}\right) \Sigma^{-} \left(V_{u1\bullet}, \ldots, V_{ut\bullet}\right)^{T} = T_{Mu},$$

say, in which Σ^{-} is a generalised inverse of Σ. The Moore–Penrose inverse is one such and is available using the `ginv()` command from the package MASS in ⓡ. Some properties are given in Appendix A.2. Of course, the test statistic $\left(V_{11\bullet}, \ldots, V_{1t\bullet}\right) \Sigma^{-} \left(V_{11\bullet}, \ldots, V_{1t\bullet}\right)^{T} = T_{M1}$ gives an unconditional assessment of mean score differences in treatments in contrast with S_{MS}, which gives a conditional assessment of mean score differences in treatments. The T_{Mu} are all asymptotically χ_{t-1}^2 distributed.

Note that the orthonormal polynomials may differ between strata, so the $V_{ui\bullet}$ are not calculated on the marginal table of counts $\{N_{ih\bullet}\}$. It does not appear possible to give a partition of an appropriate Pearson-type statistic using the T_{Mu}.

The tests for overall and even average association may have quite large degrees of freedom. This means they are seeking to detect very general alternatives to the null hypothesis of no association. The test for average association is more focused than that for overall association, but may well still have low power when $(t-1)(c-1)$ is large: the alternative to the null hypothesis is quite general. The moment tests in this section, and the generalised correlation tests in the next, offer more focused tests for alternatives that are important and easy to understand.

6.2.1 RBD Example

We now apply the mechanics earlier for the derivation of T_{M1} to the RBD. In doing so we shall meet a problem that in another context will be explored in Section 6.4. We consider only the case where there are no ties on every block.

For the RBD every treatment is observed precisely once on each block: $n_{i\bullet j} = 1$ for all i and j. Consequently $n_{\bullet\bullet j} = t$, $n_{i\bullet\bullet} = b$, $n_{\bullet\bullet\bullet} = bt$

and $p_{i\bullet j} = \frac{n_{i\bullet\bullet}}{n_{\bullet\bullet\bullet}} = \frac{1}{t}$. Hence $f_j = \left(\sqrt{n_{1\bullet j}}, \ldots, \sqrt{n_{t\bullet j}}\right)^{\mathrm{T}} = 1_t$ for all j and

$$\Sigma = \sum_{j=1}^{b} \left(I_t - \frac{f_j f_j^{\mathrm{T}}}{n_{\bullet\bullet j}}\right) = b\left(I_t - \frac{1_t 1_t^{\mathrm{T}}}{t}\right) \text{ so that } \Sigma^- = \frac{I_t - \frac{1_t 1_t^{\mathrm{T}}}{t}}{b}.$$

We need the orthonormal polynomials $\{a_{uj}(h)\}$ on the jth stratum. Since there are no ties the weight function is $\frac{1}{t}$ for $h = 1, \ldots, t$. This distribution has mean $\mu = \frac{t+1}{2}$ and variance $\sigma^2 = \frac{t^2-1}{12}$ so that $a_{1j}(h) = \frac{h-\mu}{\sigma}$ for all h. Thus

$$V_{1i\bullet} = \sum_{h=1}^{c}\sum_{j=1}^{b} N_{ihj} a_{1j}(h) = \sum_{h=1}^{c}\sum_{j=1}^{b} N_{ihj} \frac{h-\mu}{\sigma}$$

$$= \sum_{h=1}^{c} N_{ih\bullet} \frac{h-\mu}{\sigma}$$

$$= \frac{R_i - b\mu}{\sigma}.$$

Now $\left(V_{11\bullet}, \ldots, V_{1t\bullet}\right) 1_t = \frac{R_\bullet - bt\mu}{\sigma} = 0$ so that $\left(V_{11\bullet}, \ldots, V_{1t\bullet}\right)\left(I_t - \frac{1_t 1_t^{\mathrm{T}}}{t}\right) = \left(V_{11\bullet}, \ldots, V_{1t\bullet}\right)$ and $T_{\mathrm{M}1} = \left(V_{11\bullet}, \ldots, V_{1t\bullet}\right)\Sigma^-\left(V_{11\bullet}, \ldots, V_{1t\bullet}\right)^{\mathrm{T}} = \sum_{i=1}^{t} V_{1i\bullet}^2 = \sum_{i=1}^{t} \frac{(R_i - b\mu)^2}{b\sigma^2}$. We have, on recalling that

$$S = \frac{12}{bt(t+1)} \sum_{i=1}^{t}\left(R_i - \frac{b(t+1)}{2}\right)^2, \quad T_{\mathrm{M}1} = \frac{t}{t-1}S.$$

There is an apparent contradiction here, because both S and $T_{\mathrm{M}1}$ purportedly have the same asymptotic χ_{t-1}^2 distribution, yet they differ by only a constant. The brief explanation is that the tests are based on different models. The mechanics rely on theorems in Rayner and Best (2001, Chapter 4) for an unconditional model; whereas, as discussed in Section 4.5, the model for the Friedman test is conditional. We return to this issue in Section 6.5.

Before proceeding to unconditional bivariate moment tests, we need to extend the notion of correlation, which we do now.

6.3 Generalised Correlations

Generalised correlations may be defined for multivariate random variables with quite general distributions, but here for clarity we will focus on discrete bivariate and trivariate random variables.

6.3.1 Bivariate Generalised Correlations

For bivariate discrete random variables (X, Y) with $\mathrm{P}\left(X = x_i, Y = y_j\right) = p_{ij}$ for $i = 1, \ldots, I$ and $j = 1, \ldots, J$, suppose that the $a_u(X)$ are orthonormal polynomials on $\{p_{i\bullet}\}$ and the $b_v(Y)$ are orthonormal polynomials on $\{p_{\bullet j}\}$.

Putting $v = 0$ in the orthonormality condition shows that $E\left[a_u(X)\right] = 0$ for $u = 1, 2, \ldots, I - 1$: all orthonormal polynomials have mean zero. Next, putting $u = v$ shows all these orthonormal polynomials have variance one.

A saturated model for $\left\{p_{ij}\right\}$ is

$$p_{ij} = \left[1 + \sum_{u=1}^{I-1}\sum_{v=1}^{J-1}\theta_{uv}a_u(x_i)b_v(y_j)\right]p_{i\bullet}p_{\bullet j}. \tag{6.2}$$

This means that knowledge of the θ_{uv}, the orthonormal polynomials and the marginal distributions $\left\{p_{i\bullet}\right\}$ and $\left\{p_{\bullet j}\right\}$ is sufficient to determine all the p_{ij}. Now

$$E\left[a_u(X)b_v(Y)\right] = \sum_{i=1}^{t}\sum_{j=1}^{b}a_u(x_i)b_v(y_j)p_{ij} = \sum_{i=1}^{t}\sum_{j=1}^{b}a_u(x_i)b_v(y_j)p_{i\bullet}p_{\bullet j}$$

$$+ \sum_{i=1}^{t}\sum_{j=1}^{b}\sum_{u'=1}^{I-1}\sum_{v'=1}^{J-1}a_u(x_i)b_v(y_j)\theta_{u'v'}a_{u'}(x_i)b_{v'}(y_j)p_{i\bullet}p_{\bullet j}$$

$$= E\left[a_u(X)\right]E\left[b_v(Y)\right]$$

$$+ \sum_{u'=1}^{I-1}\sum_{v'=1}^{J-1}\theta_{u'v'}\left(E\left[a_u(X)a_{u'}(X)\right]E\left[b_v(Y)b_{v'}(Y)\right]\right)$$

$$= 0 \times 0 + \sum_{u'=1}^{I-1}\sum_{v'=1}^{J-1}\theta_{u'v'}\delta_{uu'}\delta_{vv'} = \theta_{uv}.$$

For $u \geq 1$ and $v \geq 1$ the (u, v)th bivariate generalised correlation is defined to be $\theta_{uv} = E\left[a_u(X)b_v(Y)\right]$. It is indeed a correlation because, using the results earlier, the orthonormal polynomials have means zero and variances all one:

$$\text{Cov}\left(a_u(X)b_v(Y)\right) = E\left[a_u(X)b_v(Y)\right] = \text{correlation}\left(a_u(X)b_v(Y)\right).$$

The 'usual' correlation, ρ_{XY}, is the $(1, 1)$th bivariate generalised correlation. The restrictions $u \geq 1$ and $v \geq 1$ are imposed because $\theta_{00} = 1$ and $\theta_{10} = \theta_{10} = 0$, none of which are genuine correlations. They are boundary conditions and not random variables.

If X and Y are independent then $\theta_{uv} = E\left[a_u(X)b_v(Y)\right] = E\left[a_u(X)\right]$ $E\left[b_v(Y)\right] = 0$ for all u and v. Conversely, if $\theta_{uv} = 0$ for all u and v, then from (6.2), $p_{ij} = p_{i\bullet}p_{\bullet j}$ for all i and j, and X and Y are independent. So we have independence if and only if *all* the generalised correlations are zero. It is now clear why $\rho_{XY} = 0$ is necessary but not sufficient for independence.

Because the θ_{uv} are correlations in the traditional sense, $-1 \leq \theta_{uv} \leq 1$, and $\theta_{uv} = \pm 1$ if and only if $a_u(X)$ and $b_v(Y)$ are linearly related. For example, if $a_2(X)$ and $b_1(Y)$ are linearly related, then for constants c, d, and

e, $ca_2(X) + db_1(Y) = e$, and $\theta_{12} = \pm 1$. Substituting for $a_2(X)$ and $b_1(Y)$ shows that Y is a quadratic function of X. Similarly if $\theta_{31} = \pm 1$ then X is a cubic function of Y.

For a table of counts $\{N_{ij}\}$ with $N_{\bullet\bullet} = n$, the score test statistic for testing $H_{uv} : \theta_{uv} = 0$ against $K_{uv} : \theta_{uv} \neq 0$ can be shown to be V_{uv}^2 in which

$$V_{uv} = \sum_{i=1}^{I} \sum_{j=1}^{J} \frac{N_{ij} a_u(x_i) b_v(y_j)}{\sqrt{n}}.$$

See Rayner and Best (2001, section 8.2). Under the null hypothesis of independence, the V_{uv} are asymptotically independent and asymptotically standard normal. When testing the null hypothesis of independence a convenient approach is to refer a particular V_{uv}^2 to the χ_1^2 distribution. Note that when there are grouped data the $\frac{V_{uv}}{\sqrt{n}}$ are, in the traditional sense, correlations between $\{a_u(x_i)\}$ and $\{b_v(y_j)\}$.

In addressing the question of independence for a data set, the analyst will usually select a set of V_{uv} and test each for consistency with zero. This may require making an adjustment for doing several hypothesis tests on the same data set. Another issue is it isn't clear how to interpret generalised correlations beyond the (1, 1)th, (1, 2)th, and (2, 1)th, such as V_{23}. Experience with component analysis in testing goodness of fit suggests that while the V_{23} component may reflect a quadratic/cubic relationship, the significance may be an alias for a different relationship. However, even suggesting that V_{rs} reflects a degree r, degree s relationship may be a step too far. So if V_{11} is not consistent with zero, it doesn't mean that $V_{11} = \pm 1$ and there is a linear/linear relationship between the variables. That may be the case, but it is much more likely that there is a tendency for larger values of Y as X increases, a tendency sufficient to invalidate the null hypothesis.

A deeper consideration may be to fit a model to the data. An initial model may be (6.2) excluding the θ_{uv} found to be consistent with zero. This will give an idea of the complexity of the relationship between the variables. There is a considerable literature on model selection, and for some problems it will be meaningful to apply appropriate techniques from that literature. In many cases this is also a step too far. If any V_{uv} are significant then the variables are not independent, and this can be formally reported.

To test for a non-zero generalised correlation, there are four obvious options. For larger sample sizes one could rely on the asymptotic normality of V_{uv} by, for example, calculating $P(V_{uv}^2 \geq v_{uv}^2)$ using the χ_1^2 distribution in which v_{uv}^2 is the observed value of V_{uv}^2. This is essentially an equal-tails Z test. For smaller sample sizes if normality is not in doubt, then a t-test may be used. If normality is in doubt, the Wilcoxon signed-rank test can be applied. The fourth option is to use a permutation test to derive p-values. To assess normality a convenient option may be the Shapiro–Wilk test.

6.3.2 Age and Intelligence Data

Rayner and Best (2001, section 8.1) give the data in Table 6.1 relating age and intelligence.

For the ranked intelligence scores and 'natural' age scores 1, 2, 3, 4, and 5, the (1, 1), (1, 2), (2, 1), and (2, 2) generalised correlations and their t-test p-values are: $\hat{\theta}_{11} = -0.3164$ (0.2267); $\hat{\theta}_{12} = -0.6178$ (0.0152); $\hat{\theta}_{21} = 0.1734$ (0.5272); and $\hat{\theta}_{22} = -0.0312$ (0.9157).

The only generalised correlation of these data that is significant at the 0.05 level is θ_{12}, the (1, 2)th generalised correlation. This suggests there is a tendency that with increasing age, intelligence increases and then decreases. This is supported by the ℝ plot of ranked scores against age

Table 6.1 Wechsler adult intelligence scores.

		Age group		
16–19	**20–34**	**35–54**	**55–69**	**69 plus**
8.62	9.85	9.98	9.12	4.80
9.94	10.43	10.69	9.89	9.18
10.06	11.31	11.40	10.57	9.27

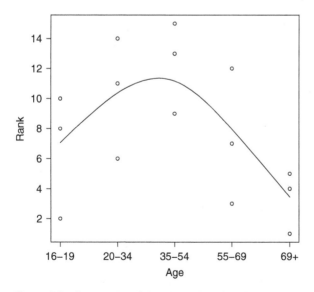

Figure 6.1 Scatter plot of the ranked intelligence scores versus age group with smooth spline highlighting the umbrella effect.

groups in Figure 6.1. The (1, 1)th generalised correlation is the Spearman correlation r_S. There is no evidence of a linear effect. If there were this would suggest that as age increases, there is a tendency for intelligence to increase, or, for that matter, to decrease.

For the (unranked) intelligence scores and natural age scores the generalised correlations and their Wilcoxon signed-rank test p-values are: $\hat{\theta}_{11} = -0.3990$ (0.1579); $\hat{\theta}_{12} = -0.5705$ (0.0067); $\hat{\theta}_{21} = 0.0549$ (0.9375); and $\hat{\theta}_{22} = -0.3280$ (0.1876). The Wilcoxon signed-rank test was employed here due to the small p-values from the Shapiro–Wilk test of normality for the (1, 1) and (1, 2)th generalised correlations.

Again, only the (1, 2)th generalised correlation is significant at the 0.05 level.

Note that testing independence using either the Pearson correlation only or the Spearman correlation only would result in a non-significant test and the conclusion that at the 0.05 level age and intelligence scores may be independent.

```
gen_cor(x = rank(score), y = age, U = 2, V = 2)
```

```
##
##          Generalised Correlations
##
## data:   rank(score), age
##
## Table of correlations and p-values
## w = 0
##                  V_11     V_12    V_21     V_22
## correlations  -0.3164  -0.6178  0.1734  -0.0312
## chi-squared    0.2204   0.0167  0.5019   0.9037
## $t$-test       0.2267   0.0152  0.5272   0.9157
## Wilcoxon       0.1307   0.0219  0.6949   0.8469
## Shapiro--Wilk  0.5870   0.5909  0.4618   0.9999
```

```
gen_cor(x = score, y = age, U = 2, V = 2)
```

```
##
##          Generalised Correlations
##
## data:   score, age
##
## Table of correlations and p-values
## w = 0
##                  V_11     V_12    V_21     V_22
## correlations  -0.3990  -0.5705  0.0549  -0.3280
## chi-squared    0.1223   0.0271  0.8317   0.2039
## $t$-test       0.2283   0.0473  0.8480   0.2438
## Wilcoxon       0.1579   0.0067  0.9375   0.1876
## Shapiro--Wilk  0.0000   0.0002  0.1656   0.7560
```

6.3.3 Trivariate Generalised Correlations

For trivariate random variables (X, Y, Z) with $P\left(X = x_i, Y = y_j, Z = z_k\right) = p_{ijk}$, three sets of orthonormal polynomials are required. Suppose that $\{a_u(X)\}$ is orthonormal on $\{p_{i\bullet\bullet}\}$, $\{b_v(Y)\}$ is orthonormal polynomials on $\{p_{\bullet j\bullet}\}$, and $\{c_w(Z)\}$ is orthonormal on $\{p_{\bullet\bullet k}\}$. A saturated model for p_{ijk} is

$$p_{ijk} = \left[1 + \sum_{u=1}^{I-1}\sum_{v=1}^{J-1}\sum_{w=1}^{K-1} \theta_{uvw} a_u\left(x_i\right) b_v\left(y_j\right) c_w\left(z_j\right)\right] p_{i\bullet\bullet} p_{\bullet j\bullet} p_{\bullet\bullet k}. \qquad (6.3)$$

This means that knowledge of the θ_{uvw}, the orthonormal polynomials and the marginal distributions $\{p_{i\bullet\bullet}\}$, $\{p_{\bullet j\bullet}\}$, and $\{p_{\bullet\bullet k}\}$ is sufficient to determine all the p_{ijk}. Provided at least two of u, v, and w are greater than zero, $\theta_{uvw} = \text{E}\left[a_u(X) b_v(Y) c_w(Z)\right]$ is defined to be the (u, v, w)th trivariate generalised correlation. In a similar manner to the bivariate case, $\theta_{000} = 1, \theta_{100} = \theta_{010} = \theta_{001} = 0$. Multivariate generalised correlations may be defined similarly. With $k > 2$ k-variate generalised correlations are no longer bounded between ±1. However, it is still true that for independence a necessary and sufficient condition is that all (u, v, w) trivariate generalised correlations are zero. Independence will fail if at least one θ_{uvw} is not consistent with zero. To assess whether or not a trivariate generalised correlation θ_{uvw} is consistent with zero for a table of counts $\{N_{ijk}\}$, the score test statistic for testing $H_{uvw} : \theta_{uvw} = 0$ against $K_{uvw} : \theta_{uvw} \neq 0$ can be shown to be V_{uvw}^2 in which, if $N_{\bullet\bullet\bullet} = n$, then

$$V_{uvw} = \sum_{i=1}^{I}\sum_{j=1}^{J}\sum_{k=1}^{K} \frac{N_{ijk} a_u\left(x_i\right) b_v\left(y_j\right) c_w\left(z_k\right)}{\sqrt{n}}.$$

Under the null hypothesis of independence the V_{uvw} are asymptotically independent and asymptotically standard normal, so the V_{uvw}^2 are asymptotically mutually independent and χ_1^2 distributed.

See Rayner and Beh (2009) for more detail on this topic.

6.3.4 Lizard Data

The data in Table 6.2 following come from Manly (2007, p. 144) and relate to the number of ants consumed by two sizes of Eastern Horned Lizards over a four month period. As size has only two levels it could be considered as either an ordered or unordered variable. The analysis here assumes the former. Thus all three margins are considered to be ordered. Natural scores 1, 2, 3, 4, and 1, 2 are used for months and size, respectively.

We now calculate the V_{uvw} in which u refers to the response, v to months, and w to size. As before only those θ_{uvw} and V_{uvw} with at least two subscripts

Table 6.2 Lizards data summarising the number of ants consumed based on month and lizard size.

	Lizard size	
Month	Small	Large
June	13, 242, 105	182, 21, 7
July	8, 59, 20	24, 312, 68
August	515, 488, 88	460, 1223, 990
September	18, 44, 21	140, 40, 27

Table 6.3 Generalised correlations $\hat{\theta}_{uv0}$ with t-test and permutation test p-values.

Response	Month (v)		
(u)	1	2	3
1	0.145/0.276/0.318	−0.453/0.023/0.015	−0.602/0.016/0.005
2	0.049/0.735/0.751	0.157/0.453/0.466	0.226/0.381/0.406
3	−0.144/0.393/0.431	0.122/0.560/0.588	0.066/0.787/0.810

positive are of interest. First, we consider bivariate generalised correlations between the data and months by taking $w = 0$. Effectively lizard size and replications are combined into new replications.

To assess whether or not the generalised correlations θ_{uv0} are consistent with zero we have the four options previously described. Table 6.3 shows the bivariate generalised correlations together with t-test and permutation test p-values. Permutation test p-values have been included due to the relatively small sample size in each cell of the data and rejection of normality for several tests.

Both the permutation test and t-test p-values tell a similar story. Both confirm that the (1, 2)th and (1, 3)th generalised sample correlations, and only these, are significantly different from zero at the 0.05 level. Thus, there is evidence, at the 0.05 level, that ants consumed and months are not independent. The analysis suggests that both quadratic and cubic month effects are required to model the bivariate data.

For $w = 1$ see the trivariate correlations, t-test, and permutation test p-values, respectively, are presented in Table 6.4. None of these generalised correlations are significantly large, and so are not required to model the data. The only generalised correlations that are significant at the 0.05 level are the (1, 2, 0)th and (1, 3, 0)th discussed previously.

Table 6.4 Genuine trivariate generalised correlations $\hat{\theta}_{uv1}$ together with p-values obtained through t-tests and permutation testing, respectively.

Response	Month (v)		
(u)	1	2	3
1	0.125/0.351/0.402	−0.257/0.215/0.248	−0.210/0.426/0.480
2	0.007/0.962/0.964	−0.100/0.635/0.649	−0.424/0.092/0.100
3	0.118/0.486/0.523	−0.048/0.819/0.831	0.204/0.400/0.437

```
gen_cor(x = ants, y = month, z = size, U = 3, V = 3, W = 1,
        n_perms = 100000)


##
##          Generalised Correlations
##
## data:  ants, month, and size
##
## Table of correlations and p-values
## w = 0
##                  V_11      V_12      V_13      V_21      V_22      V_23
## correlations   0.1451   -0.4527   -0.6023    0.0494    0.1571    0.2255
## chi-squared    0.4772    0.0266    0.0032    0.8088    0.4415    0.2693
## $t$-test       0.2756    0.0230    0.0158    0.7350    0.4533    0.3815
## Wilcoxon       0.1936    0.0164    0.0138    0.9658    0.4389    0.1074
## Permutation    0.3177    0.0153    0.0049    0.7515    0.4653    0.4072
## Shapiro--Wilk  0.2791    0.0012    0.0001    0.1950    0.1574    0.0011
##                  V_31      V_32      V_33
## correlations  -0.1443    0.1223    0.0661
## chi-squared    0.4796    0.5490    0.7461
## $t$-test       0.3934    0.5602    0.7868
## Wilcoxon       0.4489    0.7048    0.8553
## Permutation    0.4286    0.5863    0.8088
## Shapiro--Wilk  0.6649    0.1634    0.0006
##
## w = 1
##                  V_11      V_12      V_13      V_21      V_22      V_23
## correlations   0.1245   -0.2567   -0.2100    0.0069   -0.0998   -0.4237
## chi-squared    0.5420    0.2085    0.3036    0.9730    0.6248    0.0379
## $t$-test       0.3514    0.2154    0.4264    0.9623    0.6350    0.0925
## Wilcoxon       0.5646    0.4320    0.8639    0.7898    0.8303    0.0520
## Permutation    0.4023    0.2481    0.4799    0.9636    0.6491    0.0999
## Shapiro--Wilk  0.3479    0.0014    0.0001    0.5110    0.3943    0.0045
##                  V_31      V_32      V_33
## correlations   0.1183   -0.0481    0.2043
## chi-squared    0.5624    0.8136    0.3169
## $t$-test       0.4855    0.8193    0.3999
## Wilcoxon       0.4732    0.5203    0.2897
## Permutation    0.5233    0.8312    0.4371
## Shapiro--Wilk  0.8995    0.1030    0.0007
```

6.4 Unconditional Bivariate Moment Tests

Here, as in Section 6.2, a product multinomial model is assumed for the counts $\{N_{ihj}\}$ in each stratum. Within each stratum it is assumed that both the treatments and the outcome categories are ordered and scored, and so results for doubly ordered two-way tables of counts $\{N_{ij}\}$ are needed. For an unconditional model these are explored in Rayner and Best (2001, section 8.2). However, the results there assume a single multinomial: that $\{N_{ij}\}$ follows a multinomial distribution with total count $n = N_{\bullet\bullet}$ and cell probabilities $\{p_{ij}\}$ with $p_{\bullet\bullet} = 1$. However, if for each i, $i = 1, \ldots, t$, $\{N_{ij}\}$ follows a multinomial distribution with total count $n_{i\bullet}$ and cell probabilities $\{p_{ij}\}$ with $p_{i\bullet} = 1$, the likelihood for the product multinomial model differs from that for the single multinomial model only by a constant. Thus the results in Rayner and Best (2001, section 8.2) hold in this slightly different setting. In particular if $\{a_r(i)\}$ are orthonormal polynomials on the $\{p_{i\bullet}\}$ with $a_0(i) = 1$ for all i, $\{b_s(j)\}$ are orthonormal on the $\{p_{\bullet j}\}$ with $b_0(j) = 1$ for all j and if $V_{rs} = \sum_{i=1}^{t} \sum_{j=1}^{b} \frac{N_{ij} a_r(i) b_s(j)}{\sqrt{n}}$, then the $\{V_{rs}^2\}$ numerically partition the Pearson statistic and are asymptotically χ_1^2 distributed test statistics. In particular $\frac{V_{11}}{\sqrt{n}}$ is the Pearson product moment correlation for grouped data, and, if the scores are ranks, $\frac{V_{11}}{\sqrt{n}}$ is the Spearman correlation.

To use these results in the CMH context, first define, on the jth stratum, orthonormal polynomials $\{a_{rj}(i)\}$ on $(\hat{p}_{1\bullet j}, \ldots, \hat{p}_{c\bullet j})$ with $a_0(i) = 1$ for all i, and $\{b_{sj}(h)\}$ on $(\hat{p}_{\bullet 1j}, \ldots, \hat{p}_{\bullet cj})$ with $b_0(j) = 1$ for all h. Next define, for $r = 1, \ldots, t-1$ and $s = 1, \ldots, c-1$

$$V_{rsj} = \sum_{i=1}^{t} \sum_{h=1}^{c} \frac{N_{ihj} a_{rj}(i) b_{sj}(h)}{\sqrt{n_{\bullet\bullet j}}}.$$

The results in Rayner and Best (2001, section 8.2), translated to the current context, show that

$$\sum_{r=1}^{t-1} \sum_{s=1}^{c-1} V_{rsj}^2 = X_{Pj}^2,$$

the Pearson statistic on the jth stratum. The V_{rsj}^2 may be summed over strata to give an overall (r, s) correlation assessment. Because strata are independent $\sum_{j=1}^{b} V_{rsj}^2$ will have an asymptotic χ_b^2 distribution. In particular $\sum_{j=1}^{b} V_{11j}^2$ will give an overall unconditional assessment of the usual (that is, the $(1, 1)$th generalised) correlation. Since the sum of the strata Pearson statistics is the overall Pearson statistic, the $\sum_{j=1}^{b} V_{rsj}^2$ numerically partition the overall Pearson statistic.

To give an average correlation assessment, the V_{rsj} may be summed over strata to give an average (r, s) correlation assessment. The statistics $V_{rs\bullet}$ are asymptotically mutually independent $N(0, b)$ and an average (r, s) correlation assessment may be based on $V_{rs\bullet}$ or $V_{rs\bullet}^2$. As before the orthonormal polynomials may differ between strata, and so the $V_{rs\bullet}^2$ do not partition the Pearson statistic for the table $\{N_{ih\bullet}\}$. Nevertheless a plausible measure of general association, other than T_{GA} given previously, would be $\sum_{r=1}^{t-1} \sum_{s=1}^{c-1} \frac{V_{rs\bullet}^2}{b}$, which is distributed as $\chi_{(c-1)(t-1)}^2$. A more focused general association statistic would restrict r and s, perhaps $r + s \leq 3$, or $r + s \leq 4$. Note that $V_{11\bullet}^2$ can be considered the unconditional correlation analogue of the conditional CMH correlation statistic S_C.

As noted in Section 6.1, it is possible that in some strata some response categories have zero counts in them. It is then necessary to make some adjustments to the discussion here.

6.4.1 Homosexual Marriage Data

Recall that after assigning scores of 1, 2, and 3 for the responses agree, neutral, and disagree, respectively, to the proposition 'Homosexuals should be able to marry', and 1, 2, and 3 for the religions fundamentalist, moderate and liberal, respectively, Agresti (2002) found the CMH GA test statistic takes the value 19.76 with χ_4^2 p-value 0.0006, the CMH mean score test statistic takes the value 17.94 with χ_2^2 p-value 0.0001, while the CMH correlation test statistic takes the value 16.83 with χ_1^2 p-value less than 0.0001.

For these data T_{OPA} takes the value 27.09 with χ_8^2 p-value 0.0007. In calculating the latter we find the Pearson X^2 for School is 3.51 with χ_4^2 p-value 0.4760 while the Pearson X^2 for College is 23.58 with χ_4^2 p-value 0.0001. From these Pearson X^2 statistics the conditional CMH OPA test statistic S_{OPA} takes the value 26.71 with χ_8^2 p-value 0.0008. This is not given by Agresti (2002). There is strong evidence of overall association, and this is primarily due to the stratum College.

We also find $T_{GA} = 20.68$ with χ_4^2 p-value 0.0004. We previously suggested using an aggregation of $V_{rs\bullet}$ to give an unconditional general association statistic. Here $\sum_{r=1}^{t-1} \sum_{s=1}^{c-1} \frac{V_{rs\bullet}^2}{b} = 21.14$ with χ_4^2 p-value 0.0003.

Using the moment tests we find T_{M1} takes, the value 23.71 while T_{M2} takes the value 2.40; the corresponding χ_2^2 p-values being 0.0000 and 0.3017, respectively. There is strong evidence of a mean difference in religions but no evidence of a second degree effect.

The unconditional correlations, with χ_1^2 p-values in parentheses, are

	School	College
	$\hat{\theta}_{11} = -0.2019\ (0.1178),$	$\hat{\theta}_{11} = -0.5090\ (0.0000),$
	$\hat{\theta}_{12} = -0.0271\ (0.8339),$	$\hat{\theta}_{12} = 0.2122\ (0.0699),$
	$\hat{\theta}_{21} = -0.1282\ (0.3206),$	$\hat{\theta}_{21} = -0.1374\ (0.2404),$
	$\hat{\theta}_{22} = -0.0242\ (0.8511),$	$\hat{\theta}_{22} = -0.0046\ (0.9688).$

Here the first stratum is School. In the stratum College there is strong evidence of a linear-linear effect: as religion becomes increasingly liberal there is greater agreement with the proposition.

For average assessments, note that $v_{11\bullet} = -5.9132$, $v_{12\bullet} = 1.6031$, $v_{21\bullet} = -2.1672$, and $v_{22\bullet} = -0.2268$ and these are asymptotically $N(0, 2)$. The corresponding χ_1^2 p-values are 0.0000, 0.1254, 0.2570, and 0.8726. Averaged over strata, the only significant generalised correlation is the (1, 1)th. The very strong College effect dominates the weaker School effect. Also $\frac{V_{11\bullet}^2}{2} = T_C$, the unconditional correlation analogue of the conditional CMH correlation statistic S_C, takes the value 17.4830 with χ_1^2 p-value 0.0000. Here the other $V_{rs\bullet}^2$ have large p-values, reflecting no corresponding effects.

We noted in Section 2.5.5 that the CMH C statistic was not invariant when standardising within strata. However, the generalised correlations discussed in this section are genuine correlations and as such they are invariant when standardising.

For the convenience of readers the p-values for the homosexual marriage opinions example have been collected in Table 6.5. It seems clear that the dominating effect is a strong linear-linear correlation in the stratum College; the more liberal the religion the more support there is for the proposition.

Table 6.5 Homosexual marriage opinions p-values.

Conditional			Unconditional		
Statistic	**Value**	**p-value**	**Statistic**	**Value**	**p-value**
S_{OPA}	26.71	0.0008	T_{OPA}	27.09	0.0007
S_{GA}	19.76	0.0006	T_{GA}	20.68	0.0004
S_{MS}	17.94	0.0001	T_{M1}	23.71	0.0000
—	—	—	T_{M2}	2.40	0.3017
S_C	16.83	0.0000	$T_C = \frac{V_{11\bullet}^2}{2}$	17.48	0.0000
—	—	—	$\frac{V_{12\bullet}^2}{2}$	2.35	0.1254
—	—	—	$\frac{V_{21\bullet}^2}{2}$	1.29	0.2570
—	—	—	$\frac{V_{22\bullet}^2}{2}$	0.03	0.8726

As anticipated previously, for these data there is a remarkable consistency between the conditional and unconditional p-values.

We should point out that statistics used here require the definition of orthonormal functions on each stratum. However, different strata may have different numbers of distinct responses. These orthonormal functions will then be defined up to different degrees. This creates difficulty in aggregating them. For example, the data may give universal information about the usual (the $(1, 1)$th generalised) correlation but information about the $(1, 2)$th generalised correlation on only some strata. Sparse data can only be expected to yield sparse information, and in this case the degree $(1, 2)$ information may be regarded as only ancillary.

```
a_ij = b_hj = matrix(rep(1:3,2),ncol=2)
CMH(treatment = religion, response = opinion, strata = education,
    a_ij = a_ij, b_hj = b_hj, cor_breakdown = F)


##
##          Cochran Mantel Haenszel Tests
##
##                                   S df    $p$-value
## Overall Partial Association 26.71  8 7.929e-04
## General Association              19.76  4 5.561e-04
## Mean Score                       17.94  2 1.269e-04
## Correlation                      16.83  1 4.082e-05


unconditional_CMH(treatment = religion, response = opinion,
                  strata = education, a_ij = a_ij, b_hj = b_hj,
                  U = 2, V = 2)


##
##          Unconditional Analogues to the CMH Tests
##
##                 T df    p-value
## T_OPA   27.09000  8 6.814e-04
## T_GA    20.68000  4 3.659e-04
## T_M1    23.71000  2 7.119e-06
## T_M2     2.39700  2 3.017e-01
## T_C11.  17.48000  1 2.899e-05
## T_C12.   2.34800  1 1.254e-01
## T_C21.   1.28500  1 2.570e-01
## T_C22.   0.02573  1 8.726e-01
```

6.5 Unconditional General Association Tests

In Sections 6.2 and 6.4 we have derived unconditional analogues of the CMH MS and C tests. In fact we were able to go much further than that, inasmuch as the traditional CMH tests only give first moment univariate and bivariate $(1, 1)$th moment analysis, and the analogues developed earlier go to higher moments. We will rectify this deficiency of the traditional CMH tests in Chapter 7. We now derive another unconditional general association test as a Wald test, and in so doing meet a paradox in the literature.

We will limit consideration to the RBD, for reasons that will become apparent. As an aside we note that Stuart (1955) and Maxwell (1970) developed tests for the RBD with two treatments and responses that fall into at least two categories. The design for the Cochran test is randomised blocks with at least two treatments and responses that fall into just two categories. Both of these are included in the RBD or what might be called the extended Stuart-Maxwell design: randomised blocks with at least two treatments and responses that fall into at least two categories.

In Section 4.8, by using the product extended hypergeometric sampling distribution and the model $P(N_{ihj} = 1) = \theta_{ih} + \beta_{hj}$, it was shown that for the extended Stuart-Maxwell design the CMH GA test is a Wald test for testing $H : \theta = 0$ against $K : \theta \neq 0$. The test statistic is

$$S_{GA} = \hat{\theta}^T \text{Cov}^- \left(\hat{\theta} \right) \hat{\theta} = b^2 \left(\frac{t-1}{t} \right) \sum_{i=1}^{t} \hat{\theta}_i^T \hat{\Sigma}^{-1} \hat{\theta}_i.$$

Here we derive the unconditional test statistic for this design by finding the Wald-type test for a coherent model. The unconditional test statistic is a simple multiple of the conditional test statistic, yet both apparently have the same asymptotic distribution. This conflict has arisen previously in the literature. Later in this section we will resolve the apparent contradiction.

Since this is an RBD, the number of treatments of type i on block j are all 1: $n_{i \bullet j} = 1$ for all i and j. However, it is no longer assumed that the numbers of responses in each response category, the $N_{\bullet hj}$, are known before collecting the data. For this unconditional test, a product multinomial model is appropriate. On block $j, j = 1, \ldots, b$, assume that for $i = 1, \ldots, i$ the $\left(N_{i1j}, \ldots, N_{icj} \right)$ are mutually independent multinomial random variables with total count 1 and cell probabilities $\left(\pi_{i1j}, \ldots, \pi_{icj} \right)$. Thus, as before, $N_{ihj} = 1$ if an observation is found to be treatment i and is assessed to be in the hth category on block j, and zero otherwise. The model is the product of ct mutually independent multinomials. For every j,

$$E\left[N_{ihj} \right] = \pi_{ihj}, \text{Var}\left(N_{ihj} \right) = \pi_{ihj} \left(1 - \pi_{ihj} \right) \text{ and Cov}\left(N_{ihj}, N_{ih'j} \right) = -\pi_{ihj} \pi_{ih'j}.$$

Moreover, as in Section 4.8, assume $P\left(N_{ihj} = 1 \right) = \theta_{ih} + \beta_{hj}$, with the parameters interpreted as previously.

Again, as in Section 4.8, by definition $\beta_{hj} = \frac{n_{\bullet hj}}{t}$ for $h = 1, \ldots, t$. Equating the sum over blocks of the (i, h)th outcomes with its expectation gives $b\hat{\theta}_{ih} = N_{ih\bullet} - \frac{n_{\bullet h\bullet}}{t}$. As previously, the latter are indeed maximum likelihood estimators.

To evaluate the Wald-type test statistic we need the covariances between the $\hat{\theta}_{ih}$ in order to give the matrix of the quadratic form that defines the test statistic. Under the null hypothesis of no treatment effects we can write

$$\text{Var}\left(N_{ih\bullet}\right) = \sum_{j=1}^{b} \text{Var}\left(N_{ihj}\right) = \sum_{j=1}^{b} \beta_{hj}\left(1 - \beta_{hj}\right) = \sigma_{hh} \text{ say,}$$

which can be estimated by $\hat{\sigma}_{hh} = \sum_{j=1}^{b} \frac{N_{\bullet hj}\left(1 - \frac{N_{\bullet hj}}{t}\right)}{t}$. Similarly under the null hypothesis

$$\text{Cov}\left(N_{ih\bullet}, N_{ih'\bullet}\right) = \text{Cov}\left(\sum_{j=1}^{b}\sum_{j'=1}^{b} N_{ihj}, N_{ih'j'}\right)$$

$$= \text{Cov}\left(\sum_{j=1}^{b} N_{ihj}, N_{ih'j}\right)$$

$$= \sum_{j=1}^{b} \beta_{hj}\beta_{h'j} = \sigma_{hh'},$$

which can be estimated by $\hat{\sigma}_{hh'} = -\sum_{j=1}^{b} \frac{N_{\bullet hj} N_{\bullet h'\bullet}}{t^2}$.

Now $bt\hat{\theta}_i h = tN_{ih\bullet} - n_{\bullet h\bullet} = (t-1)N_{ih\bullet} - \sum_{\substack{u=1\\u\neq i}}^{t} N_{uh\bullet}$. Using this and the rules for the expectation and covariance of linear combinations of random variables, under the null hypothesis

$$(bt)^2 \text{Var}\left(\hat{\theta}_{ih}\right) = (t-1)^2 \text{Var}\left(N_{ih\bullet}\right) + \sum_{\substack{u=1\\u\neq i}}^{t} N_{uh\bullet}$$

$$= (t-1)^2 \sigma_{hh} + (t-1)\sigma_{hh} = t(t-1)\sigma_{hh},$$

and

$$\text{Cov}\left(bt\hat{\theta}_{ih}, bt\hat{\theta}_{ih'}\right) = (bt)^2 \text{Cov}\left(\hat{\theta}_{ih}, \hat{\theta}_{ih'}\right)$$

$$= \text{Cov}\left((t-1)N_{ih\bullet} - \sum_{\substack{u=1\\u\neq i}}^{t} N_{uh\bullet}, (t-1)N_{ih'\bullet} - \sum_{\substack{v=1\\v\neq i}}^{t} N_{vh'\bullet}\right)$$

$$= (t-1)^2 \text{Cov}\left(N_{ih\bullet}, N_{ih'\bullet}\right) + \text{Cov}\left(\sum_{\substack{u=1\\u\neq i}}^{t} N_{uh\bullet}, \sum_{\substack{v=1\\v\neq i}}^{t} N_{vh'\bullet}\right)$$

$$- (t-1)\text{Cov}\left(\sum_{\substack{u=1\\u\neq i}}^{t} N_{uh\bullet}, N_{ih'\bullet}\right)$$

$$- (t - 1) \; \text{Cov} \left(N_{ih\bullet}, \sum_{\substack{v=1 \\ v \neq i}}^{t} N_{vh'\bullet} \right)$$

$$= (t - 1)^2 \sigma_{hh'} + (t - 1) \sigma_{hh'}$$

$$- 0 - 0 \; (\text{treatments are independent})$$

$$= t(t - 1) \sigma_{hh'}.$$

To summarise, under the null hypothesis

$$b^2 \, \text{Var} \left(\hat{\theta}_{ih} \right) = \frac{t-1}{t} \sigma_{hh} \; \text{and} \; b^2 \, \text{Cov} \left(\hat{\theta}_{ih}, \hat{\theta}_{ih'} \right) = \frac{t-1}{t} \sigma_{hh'}.$$

Aggregating these results, again noting that treatments are independent,

$$b^2 \, \text{Cov} \left(\hat{\theta}_i \right) = \left(\frac{t-1}{t} \right) \Sigma,$$

for $i = 1, \ldots, t$. Next consider different treatments: $i \neq i'$. Again since $bt\hat{\theta}_{ih} = (t-1)N_{ih\bullet} - \sum_{\substack{u=1 \\ u \neq i}}^{t} N_{uh\bullet}$,

$$\text{Cov} \left(bt\hat{\theta}_{ih}, bt\hat{\theta}_{i'h} \right) = (bt)^2 \, \text{Cov} \left(\hat{\theta}_{ih}, \hat{\theta}_{i'h} \right)$$

$$= \text{Cov} \left((t - 1)N_{ih\bullet} - \sum_{\substack{u=1 \\ u \neq i}}^{t} N_{uh\bullet}, (t - 1)N_{i'h\bullet} - \sum_{\substack{v=1 \\ v \neq i'}}^{t} N_{vh\bullet} \right)$$

$$= \text{cov} \left((t - 1)N_{ih\bullet}, (t - 1)N_{i'h\bullet} \right)$$

$$+ \text{Cov} \left(\sum_{\substack{u=1 \\ u \neq i}}^{t} N_{uh\bullet}, \sum_{\substack{v=1 \\ v \neq i'}}^{t} N_{vh\bullet} \right)$$

$$- \text{Cov} \left(\sum_{\substack{u=1 \\ u \neq i}}^{t} N_{uh\bullet}, (t - 1)N_{i'h\bullet} \right)$$

$$- \text{Cov} \left((t - 1)N_{ih\bullet}, \sum_{\substack{v=1 \\ v \neq i'}}^{t} N_{vh\bullet} \right)$$

$$= 0 + \sum_{\substack{u=1 \\ u \neq i \\ u \neq i'}}^{t} \text{Cov} \left(N_{uh\bullet}, N_{uh\bullet} \right) - (t - 1) \, \text{Cov} \left(N_{i'h\bullet}, N_{i'h\bullet} \right)$$

$$- (t - 1) \; \text{Cov} \left(N_{ih\bullet}, N_{ih\bullet} \right)$$

$$= \sigma_{hh} \left[0 + (t - 2) - 2(t - 1) \right] = -t\sigma_{hh}.$$

Similarly $(bt)^2 \, \mathrm{Cov} \left(\hat{\theta}_{ih}, \hat{\theta}_{i'h'} \right) = -t\sigma_{hh'}$. Thus

$$b^2 \, \mathrm{Cov} \left(\hat{\theta}_i, \hat{\theta}_{i'} \right) = -\frac{1}{t}\Sigma \text{ for } i \neq i'.$$

Now since $\theta = \left(\theta_1^{\mathrm{T}}, \ldots, \theta_t^{\mathrm{T}} \right)^{\mathrm{T}}$, $b^2 \, \mathrm{Cov} \left(\hat{\theta}_i \right) = \frac{t-1}{t}\Sigma$ and $\mathrm{Cov} \left(\hat{\theta}_i, \hat{\theta}_{i'} \right) = -\frac{1}{t}\Sigma$, so that

$$b^2 \, \mathrm{Cov} \left(\hat{\theta} \right) = \left(I_t - \frac{1_t 1_t^{\mathrm{T}}}{t} \right) \otimes \Sigma.$$

As in 4.8, the Moore–Penrose inverse of $\mathrm{Var} \left(\hat{\theta} \right)$ is $\mathrm{Cov}^- \left(\hat{\theta} \right) = b^2 \left(I_t - \frac{1_t 1_t^{\mathrm{T}}}{t} \right) \otimes \Sigma^{-1}$. The unconditional test statistic, W_{U}, say, is thus given by

$$W_{\mathrm{U}} = b^2 \sum_{i=1}^{t} \hat{\theta}_i^{\mathrm{T}} \hat{\Sigma}^{-1} \hat{\theta}_i.$$

For the same reasons as for the conditional test, the asymptotic distribution of W_{U} is $\chi^2_{(c-1)(t-1)}$.

The only difference between the unconditional and conditional test statistics is the presence of the factor $\frac{t-1}{t}$ in the covariance matrix for the conditional test. How can two statistics that differ by a constant have the same distribution, albeit only asymptotically?

Before answering this question, we note that for randomised blocks in which only two responses are possible, Cochran (1950) derived a statistic Q that is equal to both the CMH GA statistic and the CMH MS statistic. Using different methods Shah and Claypool (1985) derived that statistic but without the factor $\frac{t-1}{t}$. Other authors have met the same issue, but from different approaches. Our resolution reflects that the statistics have been derived as Wald tests, but for different models.

In the present context the important properties of the Wald tests are that as the number of observations tends to infinity the test statistics of these tests tend to be chi-squared distributed, and that the tests are, in a sense, optimal. For this model there are bt observations and the asymptotics require that b tends to infinity. On each block there are $c - 1$ algebraically independent β_{hj}, and therefore $b(c - 1)$ algebraically independent β_{hj} in all. Thus as b tends to infinity so do the number of parameters. This invalidates the properties of the Wald tests. Thus while these are valid tests in as much as the null hypothesis is rejected for large values of the test statistic, the asymptotic properties may not hold.

Nevertheless valid inference may be based on resampling testing. The resampling should, however, reflect the model. For the conditional test the marginal totals for each block are assumed to be known, and this is consistent with permutation testing. The unconditional model has treatment totals known but response categories unknown, and this is consistent with the product multinomial model, and hence with using the parametric bootstrap. This should be reflected in data analysis.

6.5.1 Cross-Over Clinical Data

In StatXact (2003) three treatment, three period cross-over clinical data for 11 patients are given. The three treatments were placebo, aspirin and a new drug. The data, given in Table 6.6, were binary with success and failure being the responses. This is the Cochran design where the CMH general association and mean score designs coincide.

If the data are analysed using the ANOVA F test, the treatment p-value is 0.036. Using the Shapiro–Wilk test the residuals are consistent with normality (p-value 0.351) but 0, 1 data can hardly be held to be normal. These data should be analysed as count data. The Cochran test statistic takes the value 6.222 and the corresponding unconditional test statistic $W_U = \frac{t}{t-1} S_{GA} = \frac{3}{2} \times 6.222 = 9.333$. Using the asymptotic χ_2^2 distribution the conditional test has p-value 0.045, while the unconditional test has p-value 0.009.

Applying a permutation test using the conditional test statistic a p-value of 0.059 is reported. Since this effectively uses the product extended hypergeometric distribution this should be compared with the asymptotic 0.045 p-value. Since this study is based on only 11 patients, more credence would be given to the resampling p-value. Using the conditional test, marginal non-significance at the 0.05 level is reported.

Because the conditional and unconditional test statistics differ by a constant a parametric bootstrap p-value of 0.039 for both tests is reported by Best and Rayner (2017b). However, the bootstrap procedure uses the

Table 6.6 Cross-over clinical data.

| Treatment | Patient (blocks) | | | | | | | | | | | Total |
	1	2	3	4	5	6	7	8	9	10	11	
Placebo	0	0	1	0	0	0	1	0	0	0	0	2
Asprin	1	1	0	0	0	1	0	0	0	0	1	4
New drug	1	1	1	0	1	1	1	1	0	1	0	8
Total	2	2	2	0	1	2	2	1	0	1	1	14

product Bernoulli model and so should only be compared with the asymptotic 0.009 p-value for the unconditional test. The χ^2 p-value should be discarded, both because of the small sample size and the difficulty with the asymptotic properties of the Wald-type test. Based on the parametric bootstrap p-value of 0.039 the unconditional test concludes significance at the 0.05 level.

6.5.2 Saltiness Data

We now consider the cross-classified data from Best and Rayner (2017a) given in Table 6.7. Three products, A, B, and C, were tasted by 107 consumers who gave responses 'not salty enough', 'just about right saltiness', and 'too salty', which were there scored as 1, 2, and 3. The design here is randomised blocks with $b = 107$ and $t = 3$. For example, six subjects (blocks) gave product A a score of 2, product B a score of 3, and product C a score of 3.

The ANOVA F statistic takes the value $F = 2.23$ with p-value 0.11. From Section 4.4 the CMH mean scores test statistic S_{MS} satisfies $S_{MS} = \frac{b(t-1)F}{b-1+F}$. The asymptotic distribution of S_{MS} is χ^2_{t-1}, and this leads to $S_{MS} = 4.411$ with a p-value of 0.11. There is no evidence against the null hypothesis of similar saltiness for A, B, and C. However, both the ANOVA F and CMH MS tests require the existence of scores while the former also assumes normality. As with the previous example, whatever a formal test of significance may conclude, data consisting of 1, 2, and 3 only can hardly be held to be normal. Analysis assuming the general association analysis of this section seems more appropriate.

We now interpret the responses 1, 2, and 3 as codes, not scores, and find $S_{GA} = 4.91$ with χ^2_4 p-value of 0.30 and $W_U = 7.36$ with a χ^2_4 p-value of 0.12. Using the asymptotic χ^2_4 null distribution neither the conditional test nor the unconditional test can reject the null hypothesis at the 0.05 level. However, the p-values are quite different.

Table 6.7 Saltiness scores for three products.

A score	1			2			3		
B score	1	2	3	1	2	3	1	2	3
C score: 1	24	1	4	1	1	1	2	1	2
C score: 2	1	1	1	2	21	2	1	2	1
C score: 3	2	2	6	0	2	6	2	2	16

```
CMH(treatment = product, response = scores, strata = participant,
    test_OPA = FALSE, test_MS = FALSE, test_C = FALSE)

##
##         Cochran Mantel Haenszel Tests
##
##                        S df p-value
## General Association 4.909  4  0.2968
```

For each of the 107 consumers resampling responses can be calculated as in the following. Consider the six consumers who gave A a code of one, and B and C each a code of three. For each of these consumers resampling responses can be calculated as follows.

- For the conditional test we need to use the permutation test via the product extended hypergeometric distribution with treatment totals of 1, 1, and 1 for A, B, and C, respectively, and response totals of one, zero and two for the responses 1, 2, and 3, respectively.
- For the unconditional test we need to use the bootstrap via the product multinomial distribution. Each of A, B, and C will have responses from a multinomial distribution with total one and cell probabilities $\frac{1}{3}$, 0, and $\frac{2}{3}$ for the responses 1, 2, and 3, respectively.

We obtain a permutation test p-value of 0.30, in agreement with the asymptotic with χ_4^2 p-value. However, we obtain a bootstrap p-value of 0.30, compared with the χ_4^2 p-value of 0.12.

As in the preceding example, the χ^2 p-value for the unconditional test should be discarded, both because of the small sample size and the difficulty with the asymptotic properties of the Wald-type test.

There is strong agreement between the asymptotic χ^2 and permutation test p-values of the conditional test and the parametric bootstrap p-value for the unconditional tests. These tests find no evidence of a treatment effect at the usual levels of significance.

From an alternative viewpoint, the conditional approach corresponds to the marginal strata responses being known *a priori*, while the unconditional approach corresponds to the marginal strata responses being found *a posteriori*. If finding a p-value using the parametric bootstrap is inconvenient then a data analyst who would prefer to use the unconditional approach may have to use the conditional approach as part of a preliminary investigation of the data.

6.6   Stuart's Test

In the previous section we mentioned the test of Stuart (1955), a test for treatment effects in the RBD. It warrants closer consideration because there is some ambiguity as to whether the test is conditional or unconditional.

Of course, it was earlier found that the Kruskal–Wallis and Friedman tests were CMH MS tests, and that as CMH tests, they are conditional. That is not immediately apparent.

Stuart (1955) derived a test of marginal homogeneity, constructed from a different table of counts and a different model to those previously employed here. It is of interest that the test statistic derived is equal to S_{GA}.

To see this, first recall that the design is the RBD with two treatments and at least two categorical responses. Suppose $M_{hh'j}$ counts the number of times the first treatment is assigned to category h and the second to category h' by the jth judge. Then, for example, $M_{h\bullet\bullet}$ counts the number of times the first treatment is assigned to the hth category by all judges. In Stuart (1955) a model is given for the table $\{M_{hh'\bullet}\}$ and a Wald-type statistic developed for the null hypothesis of marginal homogeneity. This tests whether or not the probability that the first treatment falls into the hth category is the same as the probability that the second treatment also falls into this category, for all categories.

Since $N_{1hj}N_{2h'j}$ counts the number of times the jth judge assigns treatment one to category h and treatment two to category h' then $M_{hh'\bullet} = \sum_{j=1}^{b} N_{1hj}N_{2h'j}$. The marginal $M_{h\bullet\bullet}$ is thus given by

$$M_{h\bullet\bullet} = \sum_{h'=1}^{c} M_{hh'\bullet} = \sum_{h'=1}^{c}\sum_{j=1}^{b} N_{1hj}N_{2h'j}$$
$$= \sum_{j=1}^{b}\sum_{h'=1}^{c} N_{1hj}N_{2h'j} = \sum_{j=1}^{b} N_{1hj} = N_{1h\bullet},$$

since $\sum_{h'=1}^{c} N_{2h'j} = N_{2\bullet j} = 1$, being the number of times the jth judge assigns the second treatment to one of the categories. Similarly $M_{\bullet h'\bullet} = N_{2h'\bullet}$. Marginal homogeneity, which is testing whether or not the distributions $\{M_{h\bullet\bullet}\}$ and $\{M_{\bullet h\bullet}\}$ are the same, is equally testing whether or not the distributions $\{N_{1h\bullet}\}$ and $\{N_{2h'\bullet}\}$ are the same. Thus Stuart's test of marginal homogeneity is testing for the same null hypothesis as the CMH GA test.

The test statistic in Stuart (1955) is based on $(M_{h\bullet\bullet} - M_{\bullet h\bullet}) = (N_{1h\bullet} - N_{2h\bullet}) = b(\hat{\theta}_1 - \hat{\theta}_2)$ since, from Section 6.5, $b\hat{\theta}_{ih} = N_{ih\bullet} - \frac{n_{\bullet h\bullet}}{t}$. The model assumed in the Stuart derivation is a single multinomial on the table $\{M_{hh'\bullet}\}$: no marginals are assumed to be known. Thus the model apparently isn't conditional. Now while the $N_{\bullet h\bullet}$ are not specified by this model there are constraints: $N_{1\bullet\bullet} = M_{\bullet\bullet\bullet} = b$, the number of times the first treatment is classified by all judges, and similarly $N_{2\bullet\bullet} = b$. These constraints cast doubt on the conclusion that the test is unconditional.

We now show that the Stuart test statistic coincides with the conditional test statistic S_{GA}.

Stuart gives the covariance matrix V of his test statistic as having elements

$$V_{hh} = M_{h\bullet\bullet} + M_{\bullet h\bullet} - 2M_{hh\bullet}$$

and

$$V_{hh'} = -M_{hh'\bullet} - M_{h'h\bullet}.$$

Since $M_{hh'\bullet} = \sum_{j=1}^{b} N_{1hj}N_{2h'j}$, $M_{h\bullet\bullet} = N_{1h\bullet}$ and $M_{\bullet h'\bullet} = N_{2h'\bullet}$,

$$V_{hh} = N_{1h\bullet} + N_{2h\bullet} - \sum_{j=1}^{b} N_{1hj}N_{2hj}$$

and

$$V_{hh'} = -\sum_{j=1}^{b} N_{1hj}N_{2h'j} - \sum_{j=1}^{b} N_{1h'j}N_{2hj}.$$

Now in Section 6.5 we defined $\hat{\Sigma} = \mathrm{diag}\left(\sum_{j=1}^{b} \frac{N_{\bullet hj}}{t}\right) - \left\{\sum_{j=1}^{b} \frac{N_{\bullet hj}N_{\bullet h'j}}{t^2}\right\}$. For $t = 2$ this gives, first,

$$\hat{\Sigma}_{hh} = \sum_{j=1}^{b} \frac{N_{\bullet hj}}{2} - \sum_{j=1}^{b} \frac{N_{\bullet hj}^2}{4} = \frac{N_{1h\bullet}}{2} + \frac{N_{2h\bullet}}{2} - \sum_{j=1}^{b} \frac{\left(N_{1hj} + N_{2hj}\right)^2}{4}$$

$$= \frac{N_{1h\bullet}}{2} + \frac{N_{2h\bullet}}{2} - \sum_{j=1}^{b} \frac{\left(N_{1hj} + 2N_{1hj}N_{2hj} + N_{2hj}\right)}{4}$$

$$\left(\text{since } N_{ihj} = 0 \text{ or } 1, N_{ihj}^2 = N_{ihj}\right)$$

$$= \frac{N_{1h\bullet}}{4} + \frac{N_{2h\bullet}}{4} - \sum_{j=1}^{b} \frac{N_{1hj}N_{2hj}}{2}$$

$$= \frac{V_{hh}}{4},$$

and second,

$$\hat{\Sigma}_{hh'} = -\sum_{j=1}^{b} \frac{N_{\bullet hj}N_{\bullet h'j}}{4} = \sum_{j=1}^{b} \frac{\left(N_{1hj} + N_{2hj}\right)\left(N_{1h'j} + N_{2h'j}\right)}{4}$$

$$= -\frac{\sum_{j=1}^{b} N_{1hj}N_{1h'j} + \sum_{j=1}^{b} N_{1hj}N_{2h'j} + \sum_{j=1}^{b} N_{1h'j}N_{2hj} + \sum_{j=1}^{b} N_{2hj}N_{2h'j}}{4}$$

$$= -\frac{\sum_{j=1}^{b} N_{1hj}N_{2h'j} + \sum_{j=1}^{b} N_{1h'j}N_{2hj}}{4}$$

$$= \frac{V_{hh'}}{4},$$

since $\sum_{j=1}^{b} N_{1hj}N_{1h'j} = \sum_{j=1}^{b} N_{2hj}N_{2h'j} = 0$: the first (and also the second) treatment cannot be assigned to both the hth and h'th categories by the jth judge. Thus $V = 4\hat{\Sigma}$.

From Section 6.5 $b\hat{\theta}_{ih} = \sum_{j=1}^{b} N_{ihj} - \hat{\beta}_{h\bullet}$, so here with $t = 2$, $2b\hat{\theta}_{1h} = 2N_{1h\bullet} - N_{\bullet h\bullet} = 2N_{1h\bullet} - \left(N_{1h\bullet} + N_{2h\bullet}\right)$ and $b\hat{\theta}_1 = N_{1\bullet} - N_{2\bullet}$. Now from Section 4.8, $\theta_{\bullet h} = 0$, so with $t = 2$, $\theta_1 = -\theta_2$. From the expression for S_{GA} from Section 4.8, again with $t = 2$,

$$
\begin{aligned}
S_{GA} &= \frac{1}{2}b^2 \left[\hat{\theta}_1^{\mathrm{T}}\hat{\Sigma}^{-1}\hat{\theta}_1 + \hat{\theta}_2^{\mathrm{T}}\hat{\Sigma}^{-1}\hat{\theta}_2\right] \\
&= \frac{1}{2}b^2 \left[\hat{\theta}_1^{\mathrm{T}}\hat{\Sigma}^{-1}\hat{\theta}_1 + \left(-\hat{\theta}_1^{\mathrm{T}}\right)\hat{\Sigma}^{-1}\left(-\hat{\theta}_1\right)\right] \\
&= b^2\hat{\theta}_1^{\mathrm{T}}\hat{\Sigma}^{-1}\hat{\theta}_1 = \left(N_{1\bullet} - N_{2\bullet}\right)^{\mathrm{T}}\left(\frac{V}{4}\right)^{-1}\left(N_{1\bullet} - N_{2\bullet}\right) \\
&= \left(N_{1\bullet} - N_{2\bullet}\right)^{\mathrm{T}}V^{-1}\left(N_{1\bullet} - N_{2\bullet}\right),
\end{aligned}
$$

which is the Stuart test statistic.

Bibliography

Best, D. J. and Rayner, J. C. W. (2017a). A note on additional statistical analysis for JAR data. *Food Quality and Preference*, 55:6–7.

Best, D. J. and Rayner, J. C. W. (2017b). A note on an unconditional alternative to Cochran's Q. *The Mathematical Scientist*, 42:101–103.

Cochran, W. G. (1950). The comparison of percentages in matched samples. *Biometrika*, 37(3/4):256–266.

Emerson, P. L. (1968). Numerical construction of orthogonal polynomials from a general recurrence formula. *Biometrics*, 24:695–701.

Manly, B. F. J. (2007). *Randomization, Bootstrap and Monte Carlo Methods in Biology*, 3rd edition. Boca Raton, FL: Chapman & Hall/CRC.

Maxwell, A. E. (1970). Comparing the classification of subjects by two independent judges. *The British Journal of Psychiatry*, 116(535):651–655.

Rayner, J. C. W. and Beh, E. J. (2009). Towards a better understanding of correlation. *Statistica Neerlandica*, 63(3):324–333.

Rayner, J. C. W. and Best, D. J. (2001). *A Contingency Table Approach to Nonparametric Testing*. Boca Raton, FL: Chapman & Hall/CRC.

Rayner, J. C. W., Thas, O., and De Boeck, B. (2008). A generalized Emerson recurrence relation. *Australian & New Zealand Journal of Statistics*, 50(3):235–240.

Shah, A. K. and Claypool, P. L. (1985). Analysis of binary data in the randomized complete block design. *Communications in Statistics-Theory and Methods*, 14(5):1175–1179.

StatXact (2003). *User Manual*, Volume 2. CYTEL Software.

Stuart, A. (1955). A test for homogeneity of the marginal distributions in a two-way classification. *Biometrika*, 42(3/4):412–416.

7

Higher Moment Extensions to the Ordinal CMH Tests

7.1 Introduction

In Chapter 6 in constructing unconditional analogues of the CMH MS and C tests, unconditional tests were also constructed for detecting higher moment effects, both univariate and bivariate. In this chapter we will construct corresponding extensions to the ordinal CMH tests so as to also detect higher moment effects.

Section 7.2 gives the extensions of the CMH MS statistic, while Section 7.3 discusses extensions of the CMH C test. In Section 7.4 the jams and homosexual marriage examples are revisited to demonstrate the application of the extended tests.

7.2 Extensions to the CMH Mean Scores Test

To construct extensions to S_{MS} consider the scores $\left\{ b_{hj} \right\}$ on the jth stratum. A common choice of scores to give a 'mean' assessment would be the natural scores $1, 2, \ldots, c$ on all strata. However, a 'dispersion' assessment could be given by choosing, on all strata, scores $1^2, 2^2, \ldots, c^2$ and similarly greater powers might be of interest: $b_{hj} = h^r$. One problem with using these scores is that for different r the test statistics are correlated. Thus the significance or not of the test for a given r may affect the significance or not of a test for some $s, r \neq s$. We now look at using different scores with the objective of having uncorrelated test statistics. To this end denote degree r scores by b_{hj}^r, $h = 1, \ldots, c$. Define the degree r score sum for treatment i on stratum j by $\sum_{h=1}^{c} b_{hj}^r N_{ihj} = M_{ij}^r$ and the overall degree r score sum for treatment i by $\sum_{j=1}^{b} M_{ij}^r = M_i^r$. We wish to define the scores so that the degree r score sums are uncorrelated with those of degree s, $r \neq s$. In this sense the mean

An Introduction to Cochran–Mantel–Haenszel Testing and Nonparametric ANOVA, First Edition. J.C.W. Rayner and G. C. Livingston Jr.
© 2023 John Wiley & Sons Ltd. Published 2023 by John Wiley & Sons Ltd.

assessment won't influence the dispersion assessment, and so on for higher degrees. Here by dispersion we don't mean variance; what we do mean will be clarified shortly.

We also wish to allow different scores b_{hj}^r on different strata. This is appropriate if, for example, the design is randomised blocks, and to apply the Friedman test the data are ranked within blocks. To extend the analysis the data analyst seeks to test for higher moment effects. However, if ties occur with different numbers of ties on different blocks, there will different numbers of categories on different blocks. If the scores are the mid-ranks, the treatment will need to accommodate scores that may vary across strata.

We first show that for appropriately chosen $\left\{ b_{hj}^r \right\}$ if $r \neq s$, then M_{ij}^r is uncorrelated with M_{ij}^s. The sense of 'appropriately chosen' that we will pursue is $\left\{ b_{hj}^r \right\}$ being orthonormal using the weight function $\{p_{hj}\}$ in which $p_{hj} = \frac{n_{\bullet hj}}{n_{\bullet\bullet j}}$. Thus $\sum_{h=1}^{c} b_{hj}^r b_{hj}^s p_{hj} = \delta_{rs}$, for $r, s = 0, 1, \ldots, c-1$ with, to avoid ambiguity, $b_{hj}^0 = 1$ for $h = 1, \ldots, c$.

Now write $D_j = \text{diag}\left(\sqrt{p_{hj}} \right)$, $u_j = \left\{ \sqrt{p_{hj}} \right\}$, and define B_j by $\left\{ B_j \right\}_{rh} = b_{hj}^r$. Then using the orthonormality it can routinely be shown that $B_j D_j^2 B_j^T = I_c$ and $B_j D_j u_j = 0$. These will be needed next.

In order to put the current derivation in context, recall the derivation of the CMH MS test in Section 2.4. There the test statistic S_{MS} was based on the degree one treatment score sums, $\sum_{h=1}^{c} \sum_{j=1}^{b} b_{hj} N_{ihj} = \sum_{h=1}^{c} \sum_{j=1}^{b} b_{hj}^1 N_{ihj}$ in the current notation. The test statistic was a quadratic form with vector the centred degree one score sums. It was essentially derived as

$$S_{MS} = V^T \text{Cov}^-(V) V^T \quad \text{in which}$$

$$V = M_{i\bullet} - \text{E}\left[M_{i\bullet}\right] \quad \text{and} \quad \text{Cov}(V) = \sum_{j=1}^{b} S_j^2 \Lambda_j \left(I_t - u_j u_j^T \right) \Lambda_j$$

where

$$M_{ij} = \sum_{h=1}^{c} b_{hj} N_{ihj}, \quad \text{E}\left[M_{ij}\right] = n_{i\bullet j} \sum_{h=1}^{c} b_{hj} \frac{n_{\bullet hj}}{n_{\bullet\bullet j}},$$

$$\left(n_{\bullet\bullet j} - 1\right) S_j^2 = n_{\bullet\bullet j} \sum_{h=1}^{c} b_{hj}^2 n_{\bullet hj} - \left(\sum_{h=1}^{c} b_{hj} n_{\bullet hj} \right)^2,$$

$$\text{and if } p_{i\bullet j} = \frac{n_{i\bullet j}}{n_{\bullet\bullet j}} \text{ then } \Lambda_j = \text{diag}\left(\sqrt{p_{i\bullet j}} \right) \text{ and } u_j = \left\{ \sqrt{p_{i\bullet j}} \right\}.$$

Section 2.4 notation has been changed slightly since D_j has been used in a different sense in this section.

In Section 2.4 it was noted that δ_{uv} is the Kronecker delta, which takes values 1 if $u = v$ and zero otherwise. Using standard distribution theory for the product extended hypergeometric distribution, $\mathrm{E}\left[N_{ihj}\right] = \frac{n_{i\bullet j}}{n_{\bullet\bullet j}}$ and the covariance between N_{ihj} and $N_{i'h'j}$ is

$$\frac{n_{i\bullet j} n_{\bullet hj} \left(\delta_{ii'} n_{\bullet\bullet j} - n_{i'\bullet j}\right) \left(\delta_{hh'} n_{\bullet\bullet j} - n_{\bullet h'j}\right)}{n_{\bullet\bullet j}^2 \left(n_{\bullet\bullet j} - 1\right)}.$$

Write $N_{ij} = \left\{N_{ihj}\right\}$. Using this covariance it follows that $\mathrm{Cov}\left(N_{ij}, N_{i'j}\right)$ is proportional to

$$\mathrm{diag}\left(p_{hj}\right) - \left\{p_{hj} p_{h'j}\right\} = \mathrm{D}_j \left(I_c - u_j u_j^{\mathrm{T}}\right) \mathrm{D}_j.$$

The vector of degree r score sums for treatment i on stratum j using orthonormal scores is $\mathrm{B}_j N_{ij} = \sum_{h=1}^c b_{hj}^r N_{ihj}$, and since $\mathrm{Cov}\left(\mathrm{B}_j N_{ij}\right) = \mathrm{B}_j \mathrm{Cov}\left(N_{ij}\right) \mathrm{B}_j^{\mathrm{T}}$ this has covariance matrix proportional to

$$\mathrm{B}_j \mathrm{D}_j \left(I_c - u_j u_j^{\mathrm{T}}\right) \mathrm{D}_j \mathrm{B}_j^{\mathrm{T}} = \mathrm{B}_j \mathrm{D}_j^2 \mathrm{B}_j^{\mathrm{T}} - \left(\mathrm{B}_j \mathrm{D}_j u_j\right)\left(\mathrm{B}_j \mathrm{D}_j u_j\right)^{\mathrm{T}} = I_c.$$

Thus for $r \neq s$, $\sum_{h=1}^c b_{hj}^r N_{ihj}$ is uncorrelated with $\sum_{h=1}^c b_{hj}^s N_{ihj}$. Although it is not needed subsequently, it follows similarly that $\mathrm{Cov}\left(\mathrm{B}_j N_{ij}, \mathrm{B}_j N_{i'j}\right)$ is also proportional to I_c, so $\sum_{h=1}^c b_{hj}^r N_{ihj}$ is uncorrelated with $\sum_{h=1}^c b_{hj}^s N_{i'hj}$. Thus, on stratum j, for $r \neq s$, the degree r score sum for treatment i is uncorrelated with the degree s score sum for treatment i', whether or not $i = i'$.

We now show that for the scores $\left\{b_{hj}^r\right\}$, if $r \neq s$, then $\sum_{j=1}^b M_{ij}^r$ is uncorrelated with $\sum_{j=1}^b M_{ij}^s$: that is, for $r \neq s$, the degree r score sum for treatment i is uncorrelated with the degree s score sum for that treatment. The appropriate covariance is

$$\mathrm{Cov}\left(\sum_{j=1}^b M_{ij}^r, \sum_{j=1}^b M_{ij}^s\right) = \sum_{j=1}^b \sum_{j'=1}^t \mathrm{Cov}\left(M_{ij}^r, M_{ij'}^s\right)$$

$$= \sum_{j=1}^b \mathrm{Cov}\left(M_{ij}^r, M_{ij}^s\right) = 0,$$

since each summand has just been shown to be zero. The penultimate equality follows because strata are independent, so that the covariance between random variables in different strata is zero. It follows that in the sense of having zero covariance (correlation), score sums of different degrees have no bearing on each other. This is the rationale for working with orthonormal systems.

A more scalar derivation of this result was given in Rayner and Best (2018).

There are many ways to construct uncorrelated score sums. The process used subsequently here is to use the orthonormal polynomials as defined in Section 6.2.

Previously the CMH MS test statistic was denoted by S_{MS}. Denote the CMH MS test statistic of degree r by S_{MSr}. This is defined as was S_{MS}, but with scores $\left\{ b^r_{hj} \right\}$ instead of $\{b_{hj}\}$. Trivially $S_{MS1} = S_{MS}$. Each of the S_{MSr} has asymptotic distribution χ^2_{t-1}, and since the vectors defining these quadratic forms, V_r, say, are asymptotically normal and uncorrelated, they are asymptotically mutually independent. Hence the sum of the quadratic forms is asymptotically distributed as $\chi^2_{(c-1)(t-1)}$.

The CMH GA test is also asymptotically distributed as $\chi^2_{(c-1)(t-1)}$. Hence both the sum of the S_{MSr} test statistics and the CMH GA test statistic are giving a general association assessment of the null hypothesis of no association, but from different perspectives. The former involves ordinal component tests that require scores, while the latter does not. As we have discussed before, rather than aggregating all the S_{MSr} tests to form an omnibus test with at best moderate power, we recommend aggregating only those of degree up to three or four.

One consequence of the discussion here arises if, when investigating the data prior to applying either the Kruskal–Wallis or Friedman tests, it appears that a test for a second degree effect may be appropriate. Since both these tests are CMH tests we can apply the S_{MS2} test, and possibly other S_{MSr} tests. One issue of interest for RBD data is that there may be more ties in some blocks than others, in which case the weight functions $\{p_{hj}\}$ will have different numbers of elements and orthonormal functions of different degrees will be required. Thus if the orthonormal polynomials are being used, a degree two orthonormal polynomial requires moments up to the fourth. If on some block there are only four non-zero p_{hj}, it will not be possible to construct the second degree scores $\left\{ b^2_{hj} \right\}$, and hence the S_{MS2} test.

7.3 Extensions to the CMH Correlation Test

Recall the definition of the CMH C test from Section 2.5. In particular in the jth stratum, $j = 1, \ldots, b$, we needed treatment scores a_{ij}, $i = 1, \ldots, t$, and outcome scores b_{hj}, $h = 1, \ldots, c$. We defined $C_j = \sum_{i=1}^{t} \sum_{j=1}^{b} a_{ij} b_{hj} \left(N_{ihj} - E\left[N_{ihj}\right] \right)$ and $C = \sum_{j=1}^{b} C_j$, and hence $S_C = \frac{C^2}{\text{Var}(C)}$. Kronecker products were used to show that

$$\text{Var}\left(C_j\right) = \frac{n_{\bullet\bullet j}}{n_{\bullet\bullet j} - 1} \left(a_j^T V_{Tj} a_j \right) \left(b_j^T V_{Cj} b_j \right)$$

in which $a_j = \left(a_{1j}, \ldots, a_{tj}\right)^T$, $b_j = \left(b_{1j}, \ldots, b_{cj}\right)^T$, $p_{i \bullet j} = \frac{n_{i \bullet j}}{n_{\bullet\bullet\bullet}}$, $p_{\bullet hj} = \frac{n_{\bullet hj}}{n_{\bullet\bullet\bullet}}$, $V_{Tj} = \text{diag}\left(p_{i \bullet j}\right) - \{p_{i \bullet j}\} \{p_{i \bullet j}\}^T$ and $V_{Cj} = \text{diag}\left(p_{\bullet hj}\right) - \{p_{\bullet hj}\} \{p_{\bullet hj}\}^T$. Finally $\text{Var}(C) = \sum_{j=1}^{b} \text{Cov}\left(C_j\right)$ because strata are mutually independent.

To construct extensions to S_C suppose that instead of a single set of outcome scores $\{b_{hj}\}$, we have c sets of scores, $\left\{ b_{hj}^{(s)} \right\}$ for $s = 0, 1, \ldots, c-1$. Moreover suppose these scores are orthonormal in the sense that for $r, s = 0, 1, \ldots, c-1$, $\sum_{h=1}^{c} b_{hj}^{(r)} b_{hj}^{(s)} \frac{n_{\bullet hj}}{n_{\bullet \bullet j}} = \delta_{rs}$ with $b_{hj}^{(0)} = 1$ for $h = 1, \ldots, c$. Similarly instead of a single set of treatment scores $\{a_{ij}\}$, suppose we have t sets of scores, $\left\{ a_{ij}^{(r)} \right\}$ and suppose these scores are orthonormal in the sense that for $r = 0, 1, \ldots, t-1$, $\sum_{i=1}^{t} a_{ij}^{(r)} a_{ij}^{(s)} \frac{n_{i \bullet j}}{n_{\bullet \bullet j}} = \delta_{rs}$ with $a_{ij}^{(0)} = 1$ for $i = 1, \ldots, t$. Note that in the definition of the CMH C statistic in this section, although normalised scores $\{a_{ij}\}$ and $\{b_{hj}\}$ are specified, it is sufficient that they be orthogonal and not normalised.

Define CMH generalised correlation (GC) statistics

$$ S_{Crs} = \frac{C_{rs}^2}{\text{Var}\left(C_{rs}\right)} $$

in which

$$ C_{rs} = \sum_{j=1}^{b} C_j^{rs} \text{ with } C_j^{rs} = \left(a_j^{(r)} \otimes b_j^{(s)} \right), $$

recalling that $N_j = \left(N_{11j}, \ldots, N_{1cj}, \ldots, N_{t1j}, \ldots, N_{tcj} \right)^{\text{T}}$. Here $r = 1, \ldots, t-1$, $s = 1, \ldots, c-1$, $a_j^{(r)} = \left(a_{1j}^{(r)}, \ldots, a_{tj}^{(r)} \right)^{\text{T}}$ and $b_j^{(s)} = \left(b_{1j}^{(s)}, \ldots, b_{cj}^{(s)} \right)^{\text{T}}$. Trivially $S_{C11} = S_C$.

The collection of the GC statistics $\{S_{Crs}\}$ will now be examined. Write

$$ M_j = \begin{pmatrix} a_j^{(1)} \otimes b_j^{(1)} \\ \vdots \\ a_j^{(r)} \otimes b_j^{(s)} \\ \vdots \\ a_j^{(t-1)} \otimes b_j^{(c-1)} \end{pmatrix}. $$

Then

$$ M_j^{\text{T}} \left(N_j - \text{E}\left[N_j\right] \right) = \begin{pmatrix} \left(a_j^{(1)} \otimes b_j^{(1)} \right)^{\text{T}} \left(N_j - \text{E}\left[N_j\right] \right) \\ \vdots \\ \left(a_j^{(r)} \otimes b_j^{(s)} \right)^{\text{T}} \left(N_j - \text{E}\left[N_j\right] \right) \\ \vdots \\ \left(a_j^{(t-1)} \otimes b_j^{(c-1)} \right)^{\text{T}} \left(N_j - \text{E}\left[N_j\right] \right) \end{pmatrix} = \begin{pmatrix} C_j^{(11)} \\ \vdots \\ C_j^{(rs)} \\ \vdots \\ C_j^{((t-1)(c-1))} \end{pmatrix} $$

$$ = C_{Gj} $$

and $C_G = \sum_{j=1}^b C_{Gj}$. As strata are independent $\text{Var}(C_G) = \sum_{j=1}^b \text{Var}(C_{Gj})$. Note that $\text{Var}(C_{Gj})$ has typical element

$$\left(a_j^{(r)} \otimes b_j^{(s)}\right)^T \left(V_{Tj} \otimes V_{Cj}\right) \left(a_j^{(r')} \otimes b_j^{(s')}\right) = \left(a_j^{(r)^T} V_{Tj} a_j^{(r')}\right)$$
$$\otimes \left(b_j^{(s)^T} V_{Cj} b_j^{(s')}\right)$$

for $r, r' = 1, \ldots, t - 1$ and $s, s' = 1, \ldots, c - 1$. Now since $r, r' \neq 0$

$$a_j^{(r)^T} V_{Tj} a_j^{(r')} = \sum_{i=1}^t a_{ij}^{(r)} a_{ij}^{(r')} p_{i\bullet j} - \left(\sum_{i=1}^t a_{ij}^{(r)} p_{i\bullet j}\right) \left(\sum_{i=1}^t a_{ij}^{(r')} p_{i\bullet j}\right)$$

$$= \delta_{rr'} - \delta_{r0} \delta_{r'0} = \delta_{rr'}.$$

Similarly $b_j^{(s)^T} V_{Cj} b_j^{(s')} = \delta_{ss'}$. It follows that $\text{Var}(C_{Gj})$ is the $(c-1)(t-1)$ identity matrix. As $\text{Var}(C_G)$ is diagonal the S_{Crs} are uncorrelated.

The CLT assures the asymptotic multivariate normality of C_{Gj} and hence the S_{Crs}, so as the total sample size $n_{\bullet\bullet\bullet} = n_{\bullet\bullet1} + \cdots + n_{\bullet\bullet b}$ approaches infinity the S_{Crs} are asymptotically mutually independent and asymptotically have χ_1^2 distribution. The test based on this statistic could be called the CMH C_{rs} test, the CMH correlation test of degree (r, s).

From the derivation it should be clear that S_{Crs} is simply S_{C11} with the degree one orthonormal function on the $\{p_{i\bullet j}\}$ replaced by the degree r orthonormal function, and the degree one orthonormal function on the $\{p_{\bullet hj}\}$ replaced by the degree s orthonormal function. It therefore follows that with those substitutions S_{Crs} may be calculated using the method of calculation demonstrated in Section 2.5.2.

Since it is the sum of random variables that are asymptotically mutually independent, the sum of the correlation tests of all orders, $\sum_{r=1}^{t-1} \sum_{s=1}^{c-1} S_{Crs}$, is asymptotically distributed as $\chi_{(c-1)(t-1)}^2$. The CMH GA test is also asymptotically distributed as $\chi_{(c-1)(t-1)}^2$. Hence both of these tests are giving a general association assessment of the null hypothesis of no association, but from different perspectives. The former involves ordinal component tests that require scores, while the latter does not.

As we have discussed before, rather than aggregating all the CMH C_{rs} tests to form an omnibus test with at best moderate power, we recommend aggregating only those of degree $(1,1)$, $(1,2)$, $(2,1)$, and possibly also those of degree $(1,3)$, $(2,2)$, and $(3,1)$ as well. However, direct use of the extended correlation tests would normally focus on the usual correlation test, the degree $(1,1)$ test, and the tests for umbrella effects based on the degree $(1,2)$ and $(2,1)$ tests.

We observed in Section 2.5.2 that if the data consists of one stratum only then the CMH correlation statistic simplifies to

$$S_C = \frac{(n-1)\,S_{XY}^2}{S_{XX}S_{YY}} = (n-1)\,r_P^2,$$

in which r_P is the Pearson correlation coefficient. The same result holds if instead of using the first order treatment and response scores we use the degree r and degree s scores. Thus the p-values of the umbrella tests, based on the degree (1,2) and (2,1) scores, may be calculated by calculating these correlations, transforming to S_{C12} and S_{C21} and using the asymptotic χ_1^2 distribution of S_C. However, as in the previous chapter and the examples following, in practice there is no need to make the transformation.

7.4 Examples

In applications of the suite of tests discussed previously, multiple tests are being applied to a single data set. In formal hypothesis testing an adjustment would be made to the level of significance. However, we usually adopt a data analytic approach and simply report p-values for each test.

It is of value to note that the CMH tests for overall partial association and general association are omnibus tests of the null hypothesis of no association. In practice these tests often have large degrees of freedom, meaning they are seeking alternatives in a parameter space of many dimensions. More focused tests can be achieved by aggregating a limited number of the tests just discussed. We have suggested aggregating moment tests up to degree three or four and correlation tests of combined degree three or four.

We now consider two examples. The first is the jams example previously discussed in Sections 2.2, 2.5.4, 4.3, and 4.6. The second is the Homosexual Marriage example previously considered in Sections 1.4, 2.2, 2.5.3, and 3.7.

7.4.1 Jams Data

The data are JAR sweetness codes 1 to 5 assigned by eight judges and given in Table 2.2. The scores 1, 2, and 3 are assigned to the jams, and the sweetness codes are used as response scores. In Section 2.2 we found evidence of a mean effect but no evidence of a correlation effect. The scores are codes, not ranks, and are the same on each stratum.

The first issue is the construction of the orthonormal polynomials. For responses there are 24 (three jams, eight judges) equally likely responses, from which we calculate a mean and central moments up to degree six, and then use the formulae in Section 6.2. Thus since $\mu = 2.9583$, $\mu_2 = 1.3733$, and $a_1(x) = \frac{x-\mu}{\sqrt{\mu_2}}$, the values in the first column of Table 7.1 are obtained. The values for treatment are similar.

The second issue is the calculation of the test statistics. This is a randomised blocks design, and whatever the scores used the relationship between the CMH MS and ANOVA F test statistics is $S_{MS} = \frac{b(t-1)F}{b-1+F}$, and we calculate p-values using both.

The ANOVA F test statistic values and p-values for treatments (jams) using orthonormal scores of orders one, two and three are 4.6810 and 0.0278 for degree one, 1.5651 and 0.2435 for degree two and 0.6135 and 0.5554 for degree three. The corresponding values of S_{MS}, with their χ_2^2 p-values, are 6.4118 and 0.0405, 2.9237, and 0.2318, and 1.2893 and 0.5249, respectively. At the 0.05 level there is evidence of mean differences in the scores for jams, but not of higher moment effects.

Table 7.1 Jams orthonormal polynomials for (a) response and (b) treatment.

(a) Response

Response	Polynomial value		
	Degree 1	Degree 2	Degree 3
1	−1.6711	2.1605	−1.7438
2	−0.8178	−0.1364	0.9022
3	0.0356	−0.9948	−0.1212
4	0.8889	−0.4146	−1.2496
5	1.7422	1.6040	1.0817

(b) Treatment

Treatment	Polynomial value	
	Degree 1	Degree 2
1	−1.2247	0.7071
2	0	−1.4142
3	1.2247	0.7071

Table 7.2 CMH GC *p*-values when testing for zero generalised correlations for the jam data.

Response order	Treatment order	
	1	**2**
1	0.2936	0.0212
2	0.1862	0.2780
3	0.6098	0.3104

We now give the results of analysis by CMH GC statistics using two sets of treatment scores orthonormal on the treatment weights and three sets of response scores orthonormal on the response weights. The *p*-values are given in Table 7.2. Only the (1,2) *p*-value is less than 0.05. The treatment sums for jams A, B, and C are 18, 30, and 23, respectively. It seems that in passing from jam A to B and then C the sweetness is assessed to increase and then decrease.

```
# Calculating the orthogonal scores for responses
(b_hj1 = orthogonal_scores(sweetness,1))

##                  [,1]
## [1,] -1.67112762
## [2,] -0.81778585
## [3,]  0.03555591
## [4,]  0.88889767
## [5,]  1.74223943

# Calculating the orthogonal scores for treatments
(a_ij2 = orthogonal_scores(type,2))

##                 [,1]
## [1,]  0.7071068
## [2,] -1.4142136
## [3,]  0.7071068

# CMH test
CMH(treatment = type, response = sweetness, strata = judge,
        a_ij = matrix(rep(a_ij2,8),ncol = 8),
        b_hj = matrix(rep(b_hj1,8),ncol = 8),
        test_OPA = F, test_GA = F, cor_breakdown = F)


##
##          Cochran Mantel Haenszel Tests
##
```

```
##                 S df p-value
## Mean Score   6.412  2 0.04052
## Correlation 5.309   1 0.02122
```

7.4.2 Homosexual Marriage Data

For the data in Table 1.2 the CMH MS extensions of first and second degree yield p-values of 0.0001 and 0.3713, respectively. There is strong evidence of a mean difference between religions but no evidence of a second degree effect.

```
# Calculating the orthogonal scores for responses
b_hj1 = matrix(rep(orthogonal_scores(opinion,1),2), ncol = 2)
b_hj2 = matrix(rep(orthogonal_scores(opinion,2),2), ncol = 2)

# CMH test
CMH(treatment = religion, response = opinion, strata = education,
        b_hj = b_hj1, test_OPA = F, test_GA = F, test_C = F,
        cor_breakdown = F)

##
##        Cochran Mantel Haenszel Tests
##
##                 S df  p-value
## Mean Score 17.94   2 0.0001269

CMH(treatment = religion, response = opinion, strata = education,
        b_hj = b_hj2, test_OPA = F, test_GA = F, test_C = F,
        cor_breakdown = F)

##
##        Cochran Mantel Haenszel Tests
##
##                 S df p-value
## Mean Score 1.981   2  0.3713
```

In calculating the overall CMH C extensions of orders (1,1), (1,2), (2,1), and (2,2) we also calculated the generalised correlations of these orders for each stratum. See Table 7.3. The overall (1,1) correlation was identical to that found by Agresti; there is very strong evidence it is non-zero. School gave a p-value of 0.1209, while College gave a p-value of 0.0000. In the overall correlation the very strong correlation for College was not diluted by the weaker correlation for Schools. The other generalised correlations were not significant at the commonly used levels of significance.

It is worth noting here that the degree (1,2) test is assessing whether responses to the proposition increase and then decrease (or decrease then

Table 7.3 CMH generalised correlation *p*-values for the homosexual marriage data.

	CMH C extensions			
	(1,1)	**(1,2)**	**(2,1)**	**(2,2)**
School	0.1209	0.8352	0.3246	0.8523
College	0.0000	0.0718	0.2437	0.9690
Overall	0.0000	0.2902	0.1671	0.7919

increase) as religion passes from fundamentalist to moderate to liberal. The degree (2,1) test is assessing whether religion increases and then decreases (or decreases then increases) as the responses move from agree to neutral to disagree. An example of increasing and then decreasing would be passing from religion 2 to 3 and then to 1: from moderate to liberal to fundamentalist. It is harder to interpret a degree (2,2) effect and indeed higher degree correlation effects. However, all these higher degree correlations are complex correlations all of which must be zero for independence to hold.

```
# overall scores
b_hj1 = orthogonal_scores(x = opinion,degree = 1, n_strata = 2)
b_hj2 = orthogonal_scores(x = opinion,degree = 2, n_strata = 2)
a_ij1 = orthogonal_scores(x = religion,degree = 1, n_strata = 2)
a_ij2 = orthogonal_scores(x = religion,degree = 2, n_strata = 2)

# overall tests
CMH(treatment = religion, response = opinion, strata = education,
        b_hj = b_hj1, a_ij = a_ij1, test_OPA = F, test_GA = F,
        test_MS = F, cor_breakdown = F)
```

```
##
##        Cochran Mantel Haenszel Tests
##
##                   S df   p-value
## Correlation 16.83   1 4.082e-05
```

```
CMH(treatment = religion, response = opinion, strata = education,
b_hj = b_hj1, a_ij = a_ij2, test_OPA = F, test_GA = F,
test_MS = F, cor_breakdown = F)
```

```
##
##        Cochran Mantel Haenszel Tests
##
```

```
##                 S df p-value
## Correlation 1.119  1  0.2902

CMH(treatment = religion, response = opinion, strata = education,
        b_hj = b_hj2, a_ij = a_ij1, test_OPA = F, test_GA = F,
        test_MS = F, cor_breakdown = F)

##
##      Cochran Mantel Haenszel Tests
##
##                 S df p-value
## Correlation 1.909  1  0.1671

CMH(treatment = religion, response = opinion, strata = education,
        b_hj = b_hj2, a_ij = a_ij2, test_OPA = F, test_GA = F,
        test_MS = F, cor_breakdown = F)

##
##      Cochran Mantel Haenszel Tests
##
##                   S df p-value
## Correlation 0.06964  1  0.7919
```

Bibliography

Rayner, J. C. W. and Best, D. J. (2018). Extensions to the Cochran-Mantel-Haenszel mean scores and correlation tests. *Journal of Statistical Theory and Practice*, 12(3):561–574.

8

Unordered Nonparametric ANOVA

8.1 Introduction

As described here and in Chapter 10.1, nonparametric (NP) analysis of variance (ANOVA) is a suite of tests that permits a deeper scrutiny of data traditionally analysed by designs consistent with the general linear model, specifically fixed effects ANOVAs. This methodology was developed in Rayner and Best (2013), Rayner et al. (2015), in the context of multifactor ANOVA and extended to designs consistent with the general linear model in Rayner (2017). It is addressed here because it is constructed for categorical response data and gives an alternative to the ordinal Cochran–Mantel–Haenszel (CMH) tests.

The approach used in developing these tests is to construct a table of counts relating the categorical responses to the explanatory variables. Partitions of the Pearson statistic are then used to identify component random variables. Particular components may be used as data in the ANOVA of interest. Smooth product multinomial models are constructed and tests for the smooth parameters of these models are given. The smooth parameters are simply related to the ANOVA parameters showing that the tests for the smooth parameters are equivalent to tests for the ANOVA parameters. This brief outline is intended to demonstrate two things:

- First, although ANOVA responses are typically continuous, the NP ANOVA methodology assumes categorical responses.
- Second, the tests assume only weak smooth models. Assumptions such as normality of the ANOVA residuals are not necessary: inference could be based on resampling tests.

Testing is, however, very convenient if those assumptions are satisfied. When we have checked F test p-values with those obtained by resampling,

An Introduction to Cochran–Mantel–Haenszel Testing and Nonparametric ANOVA, First Edition. J.C.W. Rayner and G. C. Livingston Jr.
© 2023 John Wiley & Sons Ltd. Published 2023 by John Wiley & Sons Ltd.

we have found good agreement even when the ANOVA assumptions are questionable.

NP ANOVA treats factors for which the levels are not ordered differently from factors for which the levels are ordered. In the unordered NP ANOVA, orthonormal polynomials are constructed on the responses, and the intended ANOVA is applied to the data transformed by the first few orthonormal polynomials. These ANOVAs are, in a sense to be clarified in the following text, uncorrelated, and so the conclusions from one ANOVA do not affect the conclusions from any other. The analysis of the data transformed by the orthonormal polynomial of degree r assesses a combination of the moments up to the rth for consistency across treatments. If the first $r - 1$ analyses of a particular factor do not reject the null hypothesis of no treatment effects, then the rth analysis assesses consistency of the rth moments of that factor.

For ordered NP ANOVA, orthonormal polynomials are constructed on the responses and on the factors that have ordered levels. This results in multiple sets of orthonormal polynomials. A new response for a modified ANOVA is then constructed. The new response is the product of the uth orthonormal polynomial on the response and the vth orthonormal polynomial on the first-ordered factor ..., wth orthonormal polynomial on the last-ordered factor. The modified ANOVA is the intended ANOVA modified by omitting the factors that have ordered levels, since these have been accounted for in the new response. The conclusions for this analysis are in terms of generalised correlations, which were discussed in Section 6.3, the expectation of the product of the specified orthonormal functions. These include the $(1, 1)$th generalised correlation – the well-known in the Pearson correlation – and the $(1, 2)$th and $(2, 1)$th generalised correlations, well known as umbrella statistics. See the Strawberries example following in this section.

Even with a few treatments and a few independent variables, we are proposing performing several analyses and several tests on the same data set which would usually be regarded as exploratory data analysis. That being the case, if the assumptions underpinning the ANOVA are dubious and calculating resampling p-values is inconvenient, then the analyst can nevertheless use the ANOVA p-values and take comfort in the knowledge that ANOVA is broadly robust. Where we have calculated both, the F test and resampling p-values are generally in good agreement. It is also worth mentioning that even if all null hypotheses are true, in performing several tests at the 5% level of significance, roughly 5% of them will reject the null hypothesis. Both points should be born in mind in interpreting the testing.

Incidentally, it is of interest to note that, in a sense, the Kruskal–Wallis and Friedman tests are each equivalent to the corresponding ANOVA F and unordered NP ANOVA first degree tests.

This chapter addresses unordered NP ANOVA; ordered NP ANOVA is treated in Chapter 10. We now give an example to demonstrate the usefulness of NP ANOVA.

8.1.1 Strawberry Data

This data set, from Pearce (1960), was considered in Section 1.4 with the data given in Table 1.1. The experimental design is supplemented balanced. Pearce (1960) found a strong treatment effect: the pesticides appear to inhibit the growth of the strawberries.

The extended ANOVA ignoring order applies the general linear model platform to the data transformed by the orthonormal polynomials of degrees one, two, and so on. In general, there seems to be little to be gained from considering degree greater than three. The first degree ANOVA has pesticide effect with p-value less than 0.0001. The second degree test has a p-value of 0.0023, while the third-degree test has a p-value of 0.6148. Normality of the residuals is not in doubt for any of the three ANOVAs.

There is strong evidence of first- and second-degree effects – roughly indicating location and an unrelated (in the sense of not caused by the location effect) dispersion effect. By dispersion we mean second degree, not variance. The means for pesticides A–O are 142.4, 144.4, 103.6, 123.8, and 174.6, respectively. Although the second-degree effect reflects differences in a linear combination of both means and variances, it is natural to reflect on the variances. The variances for pesticides A–O are 455.3, 314.3, 28.8, 168.7, and 202.3, respectively. It seems the pesticides do inhibit the growth of the strawberries and that they are quite variable in how they do so.

This analysis has identified an interesting feature of the data that wasn't apparent from the traditional ANOVA. Presumably, the pesticides that least inhibits the growth of the strawberry plants, A and B, are of interest. However, these have the greatest variances: the responses to application of these pesticides will be somewhat uncertain.

It is worth emphasising that the design here is not a CMH design: the CMH methodology is not available to analyse these data. Moreover, the effects for which we have tested, of first and second degree, are uncorrelated: significance or not of one effect has not caused the significance or not of another. While NP ANOVA would usually be considered exploratory, it has revealed more than simply that there is a strong treatment effect.

8.2   Unordered NP ANOVA for the CMH Design

While CMH methods apply to three-way tables of counts, the NP ANOVA applies to these tables, and more. However since our principal focus up to this point has been CMH methods, it makes sense to introduce NP ANOVA by constructing these tests for the CMH design, and then extending to other designs. The low-degree NP ANOVA tests for these other designs are useful tests in their own right. A particular case, the Latin square design (LSD), is discussed in the next chapter. NP ANOVA is an ordinal procedure, so it is an alternative to the CMH MS and C tests. It is not an alternative to the nominal CMH tests.

To construct the new tests, it is necessary to use results about partitioning Pearson's X_P^2 statistic for three-way tables of counts. Results differ depending on whether or not a variable is ordered. In the CMH context responses are ordered, strata are not, and treatments may or may not be ordered. We defer consideration of ordered treatments to Chapter 10; in this chapter, we assume treatments are not ordered.

To better import the results needed, instead of using the notation N_{ihj} as in the CMH material, we will use N_{rij} with r for response instead of h, and with the response being the first instead of the second subscript. As previously, i reflects the treatments and j the blocks. In the singly ordered tables following, only the response is ordered. In the doubly ordered tables in Chapter 10, both response and treatments are ordered.

An outline of the procedure of the NP ANOVA for unordered treatments is as follows:

- We wish to analyse categorical data from a particular ANOVA.
- The data can be presented as a table of counts such as $\{N_{rij}\}$.
- By partitioning Pearson's X_P^2 statistic, we obtain components such as Z_{uij}.
- Suppose $a_1(r), a_2(r), \ldots$ are orthonormal polynomials on $\left\{\frac{N_{r\bullet\bullet}}{N_{\bullet\bullet\bullet}}\right\}$. For a given degree u the components $\{Z_{uij}\}$ are essentially the same as $\{a_u(r)\}$.
- Under a smooth model, the $\{a_u(r)\}$ of different degrees are uncorrelated and the degree u responses $a_u(r)$ at least asymptotically have other useful properties.
- Provided the necessary assumptions are satisfied, the ANOVA of particular interest may be applied to sets of $\{a_u(r)\}$, perhaps for $u = 1, 2, 3$. Because the $\{a_u(r)\}$ of different degrees are uncorrelated, inference for any u is not affected by inference for any other degree. The degree one

ANOVA is identical to the usual ANOVA. The degree u ANOVA gives conclusions about moments of up to the uth.

- It is desirable to confirm the assumptions of each ANOVA. Typically, we don't assess the smooth model; all that is lost if the smooth model is invalid is that it is not known if inferences of different degrees are uncorrelated. If the ANOVA assumptions are dubious, resampling p-values can be calculated. If that is not convenient, the robustness of the ANOVA suggests the p-values are generally reliable.

8.3 Singly Ordered Three-Way Tables

For the table of counts $\{N_{rij}\}$, the Pearson statistic $X_P^2 = \sum \frac{(observed-expected)^2}{expected}$ is given by

$$X_P^2 = \sum_{u=0}^{n-1} \sum_{i=1}^{t} \sum_{j=1}^{b} Z_{uij}^2,$$

in which $\{a_u(r)\}$ is a set of polynomials orthonormal on $\{p_{r\bullet\bullet}\}$ and

$$Z_{uij} = \sqrt{\frac{n}{p_{\bullet i\bullet}p_{\bullet\bullet j}}} \sum_{r=1}^{n} a_u(r) p_{rij} = \frac{n}{n_{\bullet i\bullet}n_{\bullet\bullet j}} \sum_{r=1}^{n} a_u(r) n_{rij}.$$

Here we use the notation $p_{r\bullet\bullet} = \frac{N_{r\bullet\bullet}}{n}, p_{\bullet i\bullet} = \frac{N_{\bullet i\bullet}}{n}, p_{\bullet\bullet j} = \frac{N_{\bullet\bullet j}}{n}$, and $p_{rij} = \frac{N_{rij}}{n}$. The tables are assumed to be non-trivial: all marginal sums $n_{\bullet i\bullet}$ and $n_{\bullet\bullet j}$ are positive. The expression for X_P^2 is an *arithmetic* partition of X_P^2 and requires no model. The Zs with $u = 0$, namely, Z_{0ij}, correspond to a completely unordered two-way table obtained by summing the three-way table $\{N_{rij}\}$ over rows. This table is of little interest, and henceforth, here by X_P^2, we mean $X_P^2 = \sum_{u=1}^{n-1} \sum_{i=1}^{t} \sum_{j=1}^{b} Z_{uij}^2$.

If being used to test the null hypothesis of no association against the contrary, a model is required, and then under that model X_P^2 is well known to be asymptotically distributed as $\chi^2_{(bt-1)(n-1)}$. Moreover, for each u, the statistic $\sum_{i=1}^{t} \sum_{j=1}^{b} Z_{uij}^2$ has asymptotic distribution χ^2_{bt-1}. Rayner and Beh (2009) discuss various partitions of X_P^2.

In this set-up, the N_{rij} are counts of the number of responses of type r that are treatment i on block j. A plausible initial interpretation of the partition of X_P^2 is that the Zs for a particular $u \geq 1$ give information about responses in degree u space, and this space reflects differences in the treatments in moments up to the uth.

For the table of counts $\{N_{rij}\}$, now assume the 'smooth' model that for each r the two-way table follows a multinomial distribution with total count $n_{\bullet ij}$ and probabilities, for $r = 1, \ldots, n$,

$$p_{rij} = p_{r\bullet\bullet}\left[1 + \sum_{u=1}^{n-1}\theta_{uij}a_u(r)\right].$$

The θ_{uij} are defined for $u = 1, \ldots, n-1, i = 1, \ldots, t$ and $j = 1, \ldots, b$. For each u, there are $bt - 1$ independent θ_{uij}. The null hypothesis of no association is equivalent to testing if all the θ_{uij} are zero.

Assuming the smooth model earlier, under the null hypothesis that all θ_{uij} are zero, the Zs asymptotically have mean zero, variance one, and are uncorrelated. In general, the same is true if the θ_{uij} are assumed to be of order $n^{-0.5}$. For example,

$$E\left[Z_{uij}\right] = \sqrt{\frac{n}{p_{\bullet i\bullet}p_{\bullet\bullet j}}}\sum_{r=1}^{n}a_u(r)p_{r\bullet\bullet}\left[1 + \sum_{u=1}^{n-1}\theta_{uij}a_u(r)\right]$$

$$= \sqrt{\frac{n}{p_{\bullet i\bullet}p_{\bullet\bullet j}}}\left[\sum_{r=1}^{n}a_u(r)p_{r\bullet\bullet} + \sum_{r=1}^{n}\sum_{v=1}^{n-1}a_u(r)p_{r\bullet\bullet}\theta_{vij}a_v(r)\right].$$

Now, using the orthonormality $\sum_{r=1}^{n}a_u(r)p_{r\bullet\bullet} = E\left[a_u(R)a_0(R)\right] = \delta_{u0} = 0$ for $u \geq 1$. Thus,

$$E\left[Z_{uij}\right] = \sqrt{\frac{n}{p_{\bullet i\bullet}p_{\bullet\bullet j}}}\sum_{r=1}^{n}\sum_{v=1}^{n-1}a_u(r)p_{r\bullet\bullet}\theta_{vij}a_v(r)$$

$$= \sqrt{\frac{n}{p_{\bullet i\bullet}p_{\bullet\bullet j}}}\sum_{v=1}^{n-1}\theta_{vij}\left[\sum_{r=1}^{n}a_u(r)a_v(r)p_{r\bullet\bullet}\right]$$

$$= \sqrt{\frac{n}{p_{\bullet i\bullet}p_{\bullet\bullet j}}}\sum_{v=1}^{n-1}\theta_{vij}\delta_{uv} \text{ (using the orthonormality)}$$

$$= \theta_{uij}\sqrt{\frac{n}{p_{\bullet i\bullet}p_{\bullet\bullet j}}}.$$

Details for the variances and covariances are similar to those in Rayner and Best (2013, Appendix B). These and related results for other tables are useful because if the Zs are consistent with normality, the conditions for the application of certain ANOVAs are satisfied, at least asymptotically.

Now, consider the $\{Z_{uij}\}$ for some given u. For each observation, there corresponds a triple (r, i, j). For this observation, $N_{rij} = 1$, while $N_{rij} = 0$ for all other r. Observations are considered one at a time, so multiple responses do not occur. Thus, for a given observation

$$Z_{uij} = \sqrt{\frac{n}{n_{\bullet i \bullet}} n_{\bullet \bullet j}} \sum_{r=1}^{n} a_u(r) n_{rij} \text{ (by definition)} = \sqrt{\frac{n}{n_{\bullet i \bullet} n_{\bullet \bullet j}}} a_u(r).$$

It follows that for each u, $\{Z_{uij}\} = \left\{ \sqrt{\frac{n}{n_{\bullet i \bullet} n_{\bullet \bullet j}}} a_u(r) \right\}$.

If we standardise Z_{uij} by defining $W_{uij} = \dfrac{Z_{uij}}{\sqrt{\frac{n}{n_{\bullet i \bullet} n_{\bullet \bullet j}}}}$, then since

$$E\left[Z_{uij}\right] = \theta_{uij} \sqrt{\frac{n}{p_{\bullet i \bullet} p_{\bullet \bullet j}}} = \theta_{uij} n \sqrt{\frac{n}{n_{\bullet i \bullet} n_{\bullet \bullet j}}},$$

$$E\left[W_{uij}\right] = n \theta_{uij}.$$

The W_{uij} may be used to test hypotheses about their expectations, $n\theta_{uij}$. For each u, since $Z_{uij} = \sqrt{\frac{n}{n_{\bullet i \bullet} n_{\bullet \bullet j}}} a_u(r)$, it follows that $\{W_{uij}\} = \{a_u(r)\}$. If the Ws satisfy the conditions necessary for the two-way ANOVA to be applied, the $\{a_u(r)\}$ may be used to test various null hypotheses about the expectations of the Ws.

There are $bt - 1$ independent θ_{uij}, and the same number of parameters in the standard model for the two-way ANOVA. Without specifying precisely what it is, there is a one-to-one correspondence between the elements of the two sets of parameters. Thus, the treatment test of the two-way ANOVA is testing a particular aspect of the null hypothesis of no association.

For a three-way factorial design, Rayner and Best (2013) partition the total sum of squares $\sum_{i=1}^{t} \sum_{j=1}^{b} Z_{uij}^2$ into sums of squares due to both main effects and interaction. Each sum of squares can be used to test a particular aspect of the null hypothesis of no association. When the responses are ranks, it is natural to describe these sums of squares as *analogues of the Kruskal–Wallis test*. They are well-defined test statistics, rejecting when they are large. The ANOVA test statistics divide these sums of squares by the error mean square, and these quotients are test statistics, again rejecting for large values.

The relevant point both for that design and here is that the assumptions underpinning the ANOVA aren't necessary for the ANOVA F statistics to be valid test statistics. If they weren't satisfied, then testing could proceed using resampling methods. However, if they are satisfied, then p-values can be found using the F distribution, and in fact, the entire analysis can be routinely and speedily carried out using standard ANOVA software.

Because $a_u(r)$ is a polynomial of degree u, $E\left[a_u(R)\right]$ involves moments up to the uth and since the fixed effects ANOVA tests hypotheses about the expectations of the data, the ANOVAs here are testing hypotheses about moments of the responses up to the uth.

As described in this section, NP ANOVA depends almost entirely on partitioning X_P^2 and the realisation of how the components Z_{uij} are related to the $a_u(r)$. The data may be the original data or their ranks.

8.4 The Kruskal–Wallis and Friedman Tests Are NP ANOVA Tests

We now reflect on the NP ANOVA tests when the data are ranks and the designs are the completely randomised design (CRD) and the randomised block design (RBD).

8.4.1 The Kruskal–Wallis, ANOVA F, and NP ANOVA F Tests on the Ranks Are All Equivalent

In Chapter 3, we showed that the Kruskal–Wallis test, using mid-ranks if ties occur, is equivalent to the ANOVA F test on the mid-ranks. Here we will show that the ANOVA F test, using mid-ranks if ties occur, is equivalent to the NP ANOVA F test. Thus, all three tests are equivalent.

The first-degree unordered NP ANOVA is calculated by applying the one-way ANOVA to the data $\{a_1(r)\}$. Since this ANOVA is location-scale invariant, the test is equivalent to applying the one-way ANOVA to the data $\{r\}$: the ranks. This is sufficient to establish the equivalence, but for completeness' sake, we shall show this last step algebraically.

Using the details of the parametric ANOVA for the CRD in Section 3.2, the NP ANOVA F test statistic, F_{NP}, say, is calculated using the ranks as data, so that

$$SS_{Total} = \sum_{i=1}^{t}\sum_{j=1}^{b} r_{ij}^2 - CF, \quad SS_{Treat} = \sum_{i=1}^{t} r_{i\bullet}^2 - CF, \text{ and}$$

$$SS_{Error} = SS_{Total} - SS_{Treat},$$

in which $CF = \dfrac{\left(\sum_{i=1}^{t}\sum_{j=1}^{b} r_{ij}\right)^2}{n} = \dfrac{\left[\frac{n(n+1)}{2}\right]^2}{n} = \dfrac{n(n+1)^2}{4}$ with r_{ij} being the rank of the jth replicate of treatment i. The rank sum for treatment i is $R_i = \sum_{j=1}^{b} r_{ij} = r_{i\bullet}$. Now,

$$\sum_{i=1}^{t}\sum_{j=1}^{b} r_{ij} = \frac{n(n+1)}{2},$$

$$SS_{Treat} = \sum_{i=1}^{t} \frac{R_i^2}{n_i} - CF,$$

$$SS_{\text{Total}} = \sum_{i=1}^{t} \sum_{j=1}^{b} r_{ij}^2 - CF = \sum_{i=1}^{t} \sum_{j=1}^{b} r_{ij}^2 - \frac{n(n+1)^2}{4}$$

$$= \frac{C_{\text{CRD}}(n-1)\,n\,(n+1)}{12} = C_{\text{CRD}}\sigma^2 \text{ (as in Section 3.3), and}$$

$$SS_{\text{Error}} = SS_{\text{Total}} - SS_{\text{Treat}} = C\sigma^2 - \sum_{i=1}^{t} \frac{R_i^2}{n_i} + CF,$$

in which $\sigma^2 = \frac{(n-1)n(n+1)}{12}$. The NP ANOVA F statistic is thus

$$F_{\text{NP}} = \frac{(n-t)\sum_{i=1}^{t} \frac{R_i^2}{n_i} - CF}{(t-1)\left(C\sigma^2 - \sum_{i=1}^{t} \frac{R_i^2}{n_i} + CF\right)}.$$

Now, $\frac{CF}{\sigma^2} = \frac{n(n+1)^2}{4} = \frac{3(n+1)}{n-1}$ and $\sum_{i=1}^{t} \frac{\frac{R_i^2}{n_i} - CF}{\sigma^2} = \frac{KW}{n-1}$ so

$$F_{\text{NP}} = (n-t)\,\frac{\left(\frac{\sigma^2 KW}{n-1}\right)}{(t-1)\left(C\sigma^2 - \sigma^2 \frac{KW}{n-1}\right)}$$

$$= \frac{(n-t)\,KW_{\text{A}}}{(t-1)\,(n-1-KW_{\text{A}})} = \text{ANOVA } F$$

since, as in Section 3.7, the ANOVA F statistic is $F = \frac{(n-t)W_{\text{MS}}}{(t-1)(n-1-W_{\text{MS}})}$ and as in Section 3.6, the Kruskal–Wallis tests are CMH MS tests.

8.4.2 The Friedman, ANOVA F, and NP ANOVA F Tests Are All Equivalent

In Chapter 4, we showed that the Friedman test, using mid-ranks if ties occur, is equivalent to the ANOVA F test on the mid-ranks. Here we will show that the ANOVA F test, using mid-ranks if ties occur, is equivalent to the NP ANOVA F test. Thus, all three tests are equivalent.

The first-degree unordered NP ANOVA is calculated by applying the one-way ANOVA to the data $\{a_1(r)\}$ in which the responses are the mid-ranks within blocks. Again, because the ANOVA F test is location-scale invariant, it is sufficient to apply the ANOVA to the mid-ranks within blocks. Thus, the first-degree unordered NP ANOVA is equivalent to the ANOVA F test. This is sufficient to demonstrate the equivalence of the three tests, but for completeness' sake, we shall show this last step algebraically.

Put $CF = \frac{bt(t+1)^2}{4}$ and let r_{ij} be the mid-rank of treatment i on block j. Then from Section 4.2 for the RBD ANOVA with data the mid-ranks, the factor (treatment), block, error, and total sums of squares are, respectively,

$$SS_{\text{Treat}} = \sum_{i=1}^{t} \frac{r_{i\bullet}^2}{b} - CF,$$

$$SS_{\text{Block}} = \sum_{j=1}^{b} \frac{r_{\bullet j}^2}{t} - CF,$$

$$SS_{\text{Error}} = SS_{\text{Total}} - SS_{\text{Treat}} - SS_{\text{Block}} \quad \text{and}$$

$$SS_{\text{Total}} = \sum_{i=1}^{t}\sum_{j=1}^{b} r_{ij}^2 - CF,$$

on $t-1$, $b-1$, $(b-1)(t-1)$, and $(bt-1)$ degrees of freedom, respectively. The NP ANOVA F statistic F_{NP} is calculated as follows:

$$F_{\text{NP}} = \frac{\dfrac{SS_{\text{Treat}}}{t-1}}{\dfrac{SS_{\text{Error}}}{(b-1)(t-1)}}.$$

Consider the $r_{\bullet j}$. On the jth block, if we sum ranks over all treatments, we obtain $1 + 2 + \cdots + t = \frac{t(t+1)}{2}$ if there are no ties, and the same if mid-ranks are used. The blocks sum of squares are zero, because there is no variability across blocks: all the block means are the same. (In more detail, $r_{\bullet j} = \frac{t(t+1)}{2}$, so that $\sum_{j=1}^{b} \frac{r_{\bullet j}^2}{t} = \frac{b\left[\frac{t(t+1)}{2}\right]^2}{t} = \frac{bt(t+1)^2}{4}$ and $CF = \frac{\left[\frac{bt(t+1)}{2}\right]^2}{bt} = \frac{bt(t+1)^2}{4}$, so that $SS_{\text{Block}} = \sum_{j=1}^{b} \frac{r_{\bullet j}^2}{t} - CF = 0$). Now, if $\sigma^2 = \frac{b(t-1)t(t+1)}{12}$, then

$$SS_{\text{Treat}} = \sum_{i=1}^{t} \frac{r_{i\bullet}^2}{b} - CF = \sum_{i=1}^{t} \frac{R_i^2}{b} - CF,$$

$$SS_{\text{Total}} = \sum_{i=1}^{t}\sum_{j=1}^{b} r_{ij}^2 - CF = \sum_{i=1}^{t}\sum_{j=1}^{b} r_{ij}^2 - CF = C_{\text{RBD}}\sigma^2 \text{ (as in Section 4.3)},$$

$$SS_{\text{Block}} = 0, \text{ and}$$

$$SS_{\text{Error}} = SS_{\text{Total}} - SS_{\text{Treat}} - SS_{\text{Block}} = C_{\text{RBD}}\sigma^2 - \sum_{i=1}^{t} \frac{R_i^2}{b} + CF.$$

Note from Section 4.3 that $Fdm_A = \frac{(t-1)(\sum_{i=1}^{t} R_i^2 - b\ CF)}{C_{RBD}\sigma^2}$ from which $\sum_{i=1}^{t} \frac{R_i^2}{b} - CF = \frac{C_{RBD}\sigma^2 Fdm_A}{b(t-1)}$. The NP ANOVA F statistic F_{NP} is thus

$$F_{NP} = \frac{(b-1)(t-1)\left(\sum_{i=1}^{t} \frac{R_i^2}{b} - CF\right)}{(t-1)\left(C_{RBD}\sigma^2 - \sum_{i=1}^{t} \frac{R_i^2}{b} + CF\right)}$$

$$= \frac{\dfrac{(b-1)(t-1)\,C_{RBD}\sigma^2 Fdm_A}{b(t-1)}}{(t-1)\left(C_{RBD}\sigma^2 - \dfrac{C_{RBD}\sigma^2 Fdm_A}{b(t-1)}\right)}$$

$$= \frac{(b-1)\,Fdm_A}{b(t-1) - Fdm_A} = \text{ANOVA } F,$$

from Section 4.6.

8.5 Are the CMH MS and Extensions NP ANOVA Tests?

Since the NP ANOVA is a suite of ordinal tests developed here for the CMH design, it is reasonable to ask how they relate to the CMH MS statistic and its extensions. A quick answer is that the latter have χ^2 asymptotic distributions and the former have F distributions, so they cannot be precisely the same.

While the CMH MS statistic is based on score sums, the first-degree NP ANOVA is based on $\{a_1(r)\}$, the standardised responses. In applying first-degree NP ANOVA to, for example, the homosexual marriage data, the responses in each response class are all assumed to take the value of the score for that class and aggregate these. In this sense, both procedures are based on score sums. However, they do this differently, as is evident by looking at their application in various examples.

8.5.1 Jams Data

For the data in Table 2.2, the CMH MS extensions of first- and second-degree yield p-values of 0.0405 and 0.232, respectively. The unordered NP ANOVA gave p-values of 0.028 for first degree and 0.240 for second degree. Both analyses find there is strong evidence of a mean difference between jam types but no evidence of a second degree effect: the conclusions from the two analyses are similar, but the detail is different.

```
# CMH analysis
sweetness_scores1 = orthogonal_scores(x = sweetness, degree = 1,
                                      n_strata = 8)
sweetness_scores2 = orthogonal_scores(x = sweetness, degree = 2,
                                      n_strata = 8)
CMH(treatment = type, response = sweetness, strata = judge,
    b_hj = sweetness_scores1, test_OPA = F, test_GA = F,
    test_C = F)


##
##        Cochran Mantel Haenszel Tests
##
##              S df p-value
## Mean Score 6.412  2 0.04052

CMH(treatment = type, response = sweetness, strata = judge,
    b_hj = sweetness_scores2, test_OPA = F, test_GA = F,
    test_C = F)


##
##        Cochran Mantel Haenszel Tests
##
##              S df p-value
## Mean Score 2.924  2  0.2318

# NP ANOVA analysis
np_anova(ordered_vars = sweetness,
         predictor_vars = data.frame(type,judge),uvw = 1)

## Anova Table (Type III tests)
##
## Response: ortho_poly_response
##             Sum Sq Df F value  Pr(>F)
## (Intercept) 0.07282  1  2.4736 0.13809
## type        0.27560  2  4.6810 0.02775 *
## judge       0.31226  7  1.5153 0.24031
## Residuals   0.41214 14
## ---
## Signif. codes:  0 '***' 0.001 '**' 0.01 '*' 0.05 '.' 0.1 ' ' 1

np_anova(ordered_vars = sweetness,
         predictor_vars = data.frame(type,judge),uvw = 2)

## Anova Table (Type III tests)
##
## Response: ortho_poly_response
##             Sum Sq Df F value Pr(>F)
## (Intercept) 0.02706  1  0.5919 0.4545
## type        0.14311  2  1.5651 0.2435
## judge       0.21684  7  0.6776 0.6889
## Residuals   0.64005 14
```

8.6 Extension to Other Designs

Consider data from any ANOVA consistent with the general linear model. Construct a two-way table of counts $\{N_{ri}\}$ in which r indexes the responses and i indexes sets of independent variables. For example, for a three-way factorial ANOVA, the first independent variable may be treatment A at level 1, treatment B at level 1 and replicate 1, and the next could be treatment A at level 1, treatment B at level 1 and replicate 2, and so on.

The contingency table $\{N_{ri}\}$ is a singly ordered two-way table for which Pearson's X_P^2 may be partitioned into components via $X_P^2 = \sum_{u=1}^{n-1} \sum_{i=1}^{t} Z_{ui}^2$ in which $Z_{ui} = \sum_{r=1}^{n} \frac{a_u(r)N_{ri}}{\sqrt{n}}$. In this arithmetic partition of X_P^2, $\{a_u(r)\}$ is a set of polynomials orthonormal on $\left\{\frac{N_{r\bullet}}{n}\right\}$ with $a_0(r) = 1$ for all r.

Consider now the uth orthonormal polynomial for some given u. For each observation, there corresponds a pair (r, i) and for this observation, $N_{ri} = 1$ with $N_{ri} = 0$ for all other r, so that for this pair $Z_{ui} = \frac{a_u(r)}{\sqrt{n}}$. Thus, for each u,

$$\{Z_{ui}\} = \left\{\frac{a_u(r)}{\sqrt{n}}\right\}.$$

From the orthonormality $E\left[a_u(R)\right] = 0$ for all u. It now follows that for $u \neq v$, $E\left[Z_{ui}Z_{vj}\right] = \frac{E[a_u(R)a_v(R)]}{n} = 0$. Thus, for $u \neq v$, every Z_{ui} is uncorrelated with every Z_{vj}. Moreover, as $a_u(r)$ is a polynomial in r of degree u, the non-null expectation of $a_u(R)$ involves moments up to the uth.

Whatever the ANOVA for which the data were to be analysed, the Z_{ui} are essentially the projections of the data into degree u space, and a degree u assessment of that data can be given by applying that ANOVA to the Z_{ui}. Since $\{Z_{ui}\} = \left\{\frac{a_u(r)}{\sqrt{n}}\right\}$, the same is true for $\{a_u(r)\}$. Note that with this presentation, there is no need to standardise the Zs to Ws.

An ANOVA tests hypotheses about certain parameters, and to do so a model is needed. Assume that for each response r, the table $\{N_{ri}\}$ follows a multinomial distribution with total count $n_{\bullet i}$ and probabilities $p_{ri} = p_{r\bullet}\left[1 + \sum_{u=1}^{n-1} \theta_{ui}a_u(r)\right]$ for $r = 1, \ldots, n$. The θ_{ui} are defined for $u = 1, \ldots, n-1$ and $i = 1, \ldots, n$. Under the null hypothesis of no association, all θ_{ui} are zero. The non-null expectation of $Z_{ui}\sqrt{n} = a_u(R)$ is

$$E\left[a_u(R)\right] = \sum_{r=1}^{n} a_u(r)p_{ri} = \sum_{r=1}^{n} a_u(r)p_{r\bullet}\left[1 + \sum_{v=1}^{n-1} \theta_{vi}a_v(r)\right]$$

$$= \sum_{r=1}^{n} a_u(r)p_{r\bullet} + \sum_{r=1}^{n}\sum_{v=1}^{n-1} a_u(r)p_{r\bullet}\theta_{vi}a_v(r)$$

$$= 0 + \sum_{v=1}^{n-1} \theta_{vi}\left[\sum_{r=1}^{n} a_u(r)p_{r\bullet}a_v(r)\right].$$

As before, the first term is zero because the orthonormal polynomials all have expectation zero. In the second term, the expression in brackets is δ_{uv}. The only non-zero term in the outer summation is one when $u = v$. Hence, $E\left[a_u(R)\right] = \theta_{ui}$.

Without specifying precisely what it is, there is a one-to-one correspondence between the $\left\{\theta_{ui}\right\}$ and the parameters of the ANOVA. For example, in the CRD $E\left[Z_{ui}\right] = \mu_i = \frac{\theta_{ui}}{\sqrt{n}}$.

AS $E\left[a_u(Y)\right]$ involves moments up to the uth, the degree u assessment involves moments up to the uth. These assessments are uncorrelated. If, for example, the first-degree assessment finds no significant effects, then the second-degree assessment reflects variances. Otherwise, it reflects both first and second moments.

An earlier comment was made that the tests here are based on partitioning X_P^2, and there is no difficulty with the responses being ranks: the tests are equally valid tests of aspects of the null hypothesis of no association. That some of the NP ANOVA tests coincide with rank transform tests just give an alternative *raison d'etre* for the rank transform tests.

If the responses are ranks over all observations, then $\sum_{i=1}^t Z_{1i}^2 = \sum_{i=1}^t \frac{(R_i - \mu)^2}{\sigma^2}$ could be the basis for a first-degree rank test for all fixed-effects ANOVAs.

8.6.1 Aside. The Friedman Test Revisited

For the RBD the index i could be 1 for treatment 1, block 1, 2 for treatment 1, block 2, ..., b for treatment 1, block b, $b + 1$ for treatment 2, block 1, ..., bt for treatment t, block b. The elements of $\left\{Z_{ui}\right\} = \left\{\frac{a_u(r)}{\sqrt{n}}\right\}$ run through ranks of all observations on all blocks. The RBD ANOVA can be run on the $\left\{a_u(r)\right\}$ for any given u. We have established that the unordered NP ANOVA F test for $u = 1$ is equivalent to the Friedman test, so those tests corresponding to $u = 2, 3, \ldots$ can be viewed as extensions of the Friedman test.

8.7 Latin Squares

In this section, through a single example, we consider the LSD as a particular case of the general linear model to which the NP ANOVA applies. In Chapter 9, we will consider the LSD in more detail, including 'going back to basics' and partitioning X_P^2.

8.7.1 Traffic Data

Kuehl (2000, p. 301) and Best and Rayner (2011) both consider the following scenario.

A traffic engineer conducted a study to compare the total unused red-light time for five different traffic light signal sequences. The experiment was conducted with an LSD in which blocking factors were (i) five intersections and (ii) five time of day periods. In Table 8.1, the five signal sequence treatments are shown in parentheses as A, B, C, D, E, and the numerical values are the unused red-light times in minutes.

Best and Rayner (2011) report different analyses with p-values both above and below 0.05. They also suggest the value 19.2 at intersection 2 and time period 3 might be an outlier.

A parametric analysis of the raw data gives a p-value of 0.0498 for treatments. The row effects are not significant, while the column effects give a p-value of less than 0.0001. The residuals are consistent with normality. Nevertheless, the possible outlier suggests it may be informative to consider an analysis of the ranks.

A parametric analysis of the overall ranks gives a p-value of 0.0328 for treatments. Again, the row effects are not significant, while the column effects give a p-value of less than 0.0001. The residuals are consistent with normality. The same analysis with the ranks transformed by the first orthonormal polynomial yields exactly the same results. As this is a location-scale transformation, and the ANOVA is location-scale invariant, this is as expected.

The same analysis with the ranks transformed by both the second and third degree orthonormal polynomials yields gives treatment p-values of 0.9950 and 0.3823, respectively. In both cases, the residuals are consistent with normality.

Table 8.1 Unused red-light time in minutes.

Intersection	Time period				
	1	2	3	4	5
1	15.2 (A)	33.8 (B)	13.5 (C)	27.4 (D)	29.1 (E)
2	16.5 (B)	26.5 (C)	19.2 (D)	25.8 (E)	22.7 (A)
3	12.1 (C)	31.4 (D)	17.0 (E)	31.5 (A)	30.2 (B)
4	10.7 (D)	34.2 (E)	19.5 (A)	27.2 (B)	21.6 (C)
5	14.6 (E)	31.7 (A)	16.7 (B)	26.3 (C)	23.8 (D)

It seems there is evidence of location differences in the treatments as revealed by the ranked data; there was no evidence of higher-order differences.

```
np_anova(ordered_vars = rank(minutes),
         predictor_vars = data.frame(treatment,intersection,
                                     time_of_day), uvw = 1)

## Anova Table (Type III tests)
##
## Response: ortho_poly_response
##                Sum Sq Df F value    Pr(>F)
## (Intercept)   0.07669  1 15.4163  0.002012 **
## treatment     0.07508  4  3.7732  0.032787 *
## intersection  0.02369  4  1.1907  0.364023
## time_of_day   0.84154  4 42.2938 5.575e-07 ***
## Residuals     0.05969 12
## ---
## Signif. codes:  0 '***' 0.001 '**' 0.01 '*' 0.05 '.' 0.1 ' ' 1

np_anova(ordered_vars = rank(minutes),
         predictor_vars = data.frame(treatment,intersection,
                                     time_of_day), uvw = 2)

## Anova Table (Type III tests)
##
## Response: ortho_poly_response
##                Sum Sq Df F value  Pr(>F)
## (Intercept)   0.11205  1  3.9396 0.07050.
## treatment     0.00550  4  0.0483 0.99499
## intersection  0.19480  4  1.7122 0.21174
## time_of_day   0.45838  4  4.0289 0.02683 *
## Residuals     0.34132 12
## ---
## Signif. codes:  0 '***' 0.001 '**' 0.01 '*' 0.05 '.' 0.1 ' ' 1

np_anova(ordered_vars = rank(minutes),
         predictor_vars = data.frame(treatment,intersection,
                                     time_of_day), uvw = 3)

## Anova Table (Type III tests)
##
## Response: ortho_poly_response
##                Sum Sq Df F value Pr(>F)
## (Intercept)   0.00017  1  0.0042 0.9492
## treatment     0.18238  4  1.1441 0.3823
## intersection  0.06537  4  0.4101 0.7981
## time_of_day   0.27402  4  1.7190 0.2103
## Residuals     0.47823 12
```

8.8   Balanced Incomplete Blocks

Unlike CMH methods, NP ANOVA has no difficulties with the incomplete designs, such as the Latin squares and balanced incomplete block designs. Rather than an in-depth treatment for the latter, we re-analyse the ice cream vanilla flavouring example of Section 5.6 using unordered NP ANOVA.

8.8.1   Ice Cream Flavouring Data Revisited

First recall that the standard parametric analysis on the ranks gave a treatments *p*-value of 0.0049.

As expected, the first-degree analysis was identical to the parametric analysis on the ranks. The second degree analysis had a treatments *p*-value of 0.6118. There was insufficient data for a third-degree analysis. For both analyses, the residuals are consistent with normality. There is evidence of first degree but not second-degree effects.

```
np_anova(ordered_vars = rank,
        predictor_vars = data.frame(variety, judge), uvw = 1)

## Anova Table (Type III tests)
##
## Response: ortho_poly_response
##                Sum Sq Df F value   Pr(>F)
## (Intercept) 0.03768  1  2.1099 0.184416
## variety     0.85714  6  8.0000 0.004904 **
## judge       0.19048  6  1.7778 0.221005
## Residuals   0.14286  8
## ---
## Signif. codes:  0 '***' 0.001 '**' 0.01 '*' 0.05 '.' 0.1 ' ' 1

np_anova(ordered_vars = rank,
        predictor_vars = data.frame(variety, judge), uvw = 2)

## Anova Table (Type III tests)
##
## Response: ortho_poly_response
##                Sum Sq Df F value Pr(>F)
## (Intercept) 0.02826  1  0.3573 0.5665
## variety     0.36735  6  0.7742 0.6118
## judge       0.08163  6  0.1720 0.9771
## Residuals   0.63265  8
```

Bibliography

Best, D. J. and Rayner, J. C. W. (2011). Nonparametric tests for Latin squares. NIASRA Statistics Working Paper Series, 11-11.

Kuehl, R. (2000). *Design of Experiments: Statistical Principles of Research Design and Analysis*. Belmont, CA: Duxbury Press.

Pearce, S. C. (1960). Supplemented balance. *Biometrika*, 47(3/4):263–271.

Rayner, J. C. W. (2017). Extended ANOVA. *Journal of Statistical Theory and Practice*, 11(1):208–219.

Rayner, J. C. W. and Eric J. Beh (2009). Components of Pearson's Statistic for at Least Partially Ordered m-Way Contingency Tables. *Advances in Decision Sciences*, 2009:9, Article ID 980706. https://doi.org/10.1155/2009/980706.

Rayner, J. C. W. and Best, D. J. (2013). Extended ANOVA and rank transform procedures. *Australian & New Zealand Journal of Statistics*, 55(3):305–319.

Rayner, J. C. W., Best, D. J., and Thas, O. (2015). Extended analysis of at least partially ordered multi-factor ANOVA. *Australian & New Zealand Journal of Statistics*, 57(2):211–224.

9

The Latin Square Design

9.1 Introduction

The Latin square is an efficient design to use when there are two blocking factors. Moreover, it is orthogonal and simple to analyse. It is not possible to use Cochran–Mantel–Haenszel (CMH) methods, since they require that every treatment to be present on every block or stratum. Fortunately, non-parametric ANOVA is available, but here we give a new rank test based on the treatment rank sums. We give a plausibility argument that is accessible to any who understand the derivation of the parametric ANOVA test; later a more technical argument is given. A form is given for when there are no ties and adjustments are given for when ties occur: both in general and for when mid-ranks are used.

We begin in Section 9.2 by outlining the parametric ANOVA and then, in the following section, an informal derivation of the new rank test that for brevity we call the RL test.

The Latin square design (LSD) is constructed to efficiently permit double blocking and can be expected that at least one of these will be sufficiently substantive that the treatment rankings will be affected. One way of dealing with this is to use alignment. This is discussed in Section 9.4.

In Section 9.5, we report on a simulation study comparing the performance of the RL test, the ANOVA F tests based on both the raw data and the ranks, and considering, where appropriate, both aligned and non-aligned tests. We analyse three data sets in Section 9.6. In Section 9.7, we apply the orthogonal contrasts material of Section 5.5 to the three examples discussed in Section 9.6. Finally, in Section 9.8, we derive the RL test based using the unordered non-parametric ANOVA approach.

The development in this chapter will put the LSD on a similar footing to the completely randomised design (CRD), randomised block design (RBD), and balanced incomplete block design (BIBD), discussed in previous

An Introduction to Cochran–Mantel–Haenszel Testing and Nonparametric ANOVA,
First Edition. J.C.W. Rayner and G. C. Livingston Jr.
© 2023 John Wiley & Sons Ltd. Published 2023 by John Wiley & Sons Ltd.

chapters. All have rank tests for treatment effects with simple forms for untied data and adjustments for when there are ties generally and for when there are ties and mid-ranks are used. All have competing F tests on the ranks (or aligned ranks) that have comparable power and control the type I error rates.

9.2 The Latin Square Design and Parametric Model

An LSD is the arrangement of t treatments, each one repeated t times, in such a way that each treatment appears exactly once in each row and exactly once in each column in the design. Typically, but not exclusively, the treatments are denoted by Roman characters. The following are two examples:

$$\begin{pmatrix} 1 & 2 & 3 \\ 2 & 3 & 1 \\ 3 & 1 & 2 \end{pmatrix} \text{ and } \begin{pmatrix} A & B & D & C \\ B & C & A & D \\ C & D & B & A \\ D & A & C & B \end{pmatrix}.$$

The LSD possibly represents the most popular alternative design when two blocking factors need to be controlled. It is an extreme example of an incomplete block design, with any combination of levels involving the two blocking factors assigned to one treatment only, rather than to all. A full three-way ANOVA with t treatments on both blocks would require t^3 observations, whereas the LSD needs only t^2 observations.

The classical additive model for the LSD is

$$Y_{ijk} = \mu + A_i + B_j + C_k + E_{ijk},$$

for $i = 1, \ldots, t, j = 1, \ldots, t, k = 1, \ldots, t$, in which μ is the overall mean, the A_i are treatment effects, the B_j are row effects, the C_k are column effects, and E_{ijk} are errors. In the fixed-effects model, the parameters are all constants, with constraints $\sum_{i=1}^{t} A_i = \sum_{j=1}^{t} B_j = \sum_{k=1}^{t} C_k = 0$. The errors are assumed to be normally, independently and identically distributed with zero mean and constant variance: $\text{IIN}(0, \sigma^2)$.

Put $Y_{i\bullet\bullet}$, the sum over the Latin square of the observations for treatment i, $Y_{\bullet j\bullet}$, the sum of the observations for row j, and $Y_{\bullet\bullet k}$, the sum of the observations for column k, $Y_{\bullet\bullet\bullet}$ the sum of all the observations, and $CF = \frac{Y_{\bullet\bullet\bullet}^2}{t^2}$, a correction factor. Then the treatment, row, column, total, and error sums of squares are, respectively,

$$SS_{\text{Treat}} = \sum_{i=1}^{t} \frac{Y_{i\bullet\bullet}^2}{t} - CF, \ SS_{\text{Row}} = \sum_{j=1}^{t} \frac{Y_{\bullet j\bullet}^2}{t} - CF, \ SS_{\text{Col}} = \sum_{j=1}^{t} \frac{Y_{\bullet\bullet k}^2}{t} - CF,$$

$$SS_{\text{Total}} = \sum_{\text{all cells}} Y_{ijk}^2 - CF, \text{ and } SS_{\text{Error}} = SS_{\text{Total}} - SS_{\text{Treat}} - SS_{\text{Row}} - SS_{\text{Col}}.$$

The treatment, row, and column degrees of freedom are all $t - 1$, while the error degrees of freedom are $(t - 1)(t - 2)$. The mean squares for treatments, rows, and columns are the sums of squares divided by their degrees of freedom, and the ANOVA F statistics are those mean squares divided by the error mean squares:

$$MS_{\text{Treat}} = \frac{SS_{\text{Treat}}}{t - 1} \quad \text{and} \quad MS_{\text{Error}} = \frac{SS_{\text{Error}}}{(t - 1)(t - 2)}.$$

To test the null hypothesis that $\{\alpha_i\} = 0$ against $\{\alpha_i\} \neq 0$, the ANOVA F test statistic is $F = \frac{MS_{\text{Treat}}}{MS_{\text{Error}}}$ which under the model assumptions has sampling distribution $F_{t-1,\,(t-1)(t-2)}$. Block effects may be assessed similarly, although they are not usually of interest.

9.3 The RL Test

Suppose data for a $t \times t$ LSD are ranked overall and that R_{ijk} is the rank of treatment i in row j and column k. Under the null hypothesis of no difference in treatment distributions, the mean of the ranks is $\mu = \frac{t^2+1}{2}$, and if there are no ties, the variance of the ranks is $\sigma^2 = \frac{(t^2-1)(t^2+1)}{12}$. The effect of ties will be considered subsequently.

Suppose now that the standardised ranks, $\frac{R_{ijk}-\mu}{\sigma} = Y_{ijk}$, say, are analysed using the ANOVA F test. Then $Y_{\bullet\bullet\bullet} = 0$ and the correction factor, $\frac{Y_{\bullet\bullet\bullet}^2}{t^2}$, is zero. The treatment sum of squares is $\sum_{i=1}^{t} \frac{Y_{i\bullet\bullet}^2}{t} = SS_{\text{Treat}}$. From their definition, the $Y_{i\bullet\bullet} = \frac{R_{i\bullet\bullet}-t\mu}{\sigma}$ have means zero and variances t, since there are t occurrences of treatment i in the Latin square. In Section 9.8, we will show that the $Y_{i\bullet\bullet}$ are asymptotically uncorrelated.

We know that if $X_1, \ldots, X_n$ are mutually independent $N(\mu, \tau^2)$ random variables, then $\sum_{i=1}^{n} \frac{(X_i-\overline{X})^2}{\tau^2}$ is χ_{n-1}^2 distributed. By the central limit theorem (CLT) the $Y_{i\bullet\bullet}$ are asymptotically normal and hence asymptotically mutually independent. In large samples, under the null hypothesis of equality of treatment rank means, this result may be applied to the $Y_{i\bullet\bullet}$. It follows that SS_{Treat} is then approximately χ_{t-1}^2 distributed. In terms of the treatment rank sums the test statistic is $\sum_{i=1}^{t} \frac{Y_{i\bullet\bullet}^2}{t} = \sum_{i=1}^{t} \frac{(R_{i\bullet\bullet}-t\mu)^2}{t\sigma^2} = RL$, say, and when there are no ties we have

$$RL = \frac{12 \sum_{i=1}^{t} (R_{i\bullet\bullet} - t\mu)^2}{t(t^2 - 1)(t^2 + 1)}.$$

Now $\sum_{i=1}^{t} \left(R_{i\bullet\bullet} - t\mu \right)^2 = \sum_{i=1}^{t} R_{i\bullet\bullet}^2 - t^3\mu^2$ so that

$$RL = -\frac{3t^2 \left(t^2 + 1\right)}{t^2 - 1} + \frac{12 \sum_{i=1}^{t} R_{i\bullet\bullet}^2}{t \left(t^2 - 1\right) \left(t^2 + 1\right)}.$$

Of course, for the LSDs used in practice, the number of summands, t, is hardly large, and appealing to the CLT may be considered dubious. Whether the conclusions of this approach hold is a matter for the subsequent empirical investigation.

Suppose ties occur and mid-ranks are used. If the gth group of ties is of size t_g, then by the same argument that was used in Section 3.3 an adjusted statistic is $RL_M = \frac{RL}{C_{LS}}$ in which

$$C_{LS} = 1 - \sum_{g=1}^{G} \frac{t_g^3 - t_g}{\left(t^2 - 1\right) t \left(t^2 + 1\right)}.$$

Whether or not there are ties, the variance of the ranks is $\sum_{j=1}^{t} \sum_{k=1}^{t} \frac{r_{jk}^2}{t^2} - \left(\frac{t^2+1}{2}\right)^2$ in which r_{jk} is the overall rank of the observation in row j and column k. Thus,

$$RL_A = \sum_{i=1}^{t} \frac{\left(R_{i\bullet\bullet} - t\mu\right)^2}{\sum_{j=1}^{t} \sum_{k=1}^{t} \frac{r_{jk}^2}{t} - \frac{t(t^2+1)^2}{4}} = \frac{-\frac{t^3(t^2+1)^2}{4} + \sum_{i=1}^{t} R_{i\bullet\bullet}^2}{\frac{-t(t^2+1)^2}{4} + \sum_{j=1}^{t} \sum_{k=1}^{t} \frac{r_{jk}^2}{t}}.$$

Again, this adjustment is similar in form to those for the Kruskal–Wallis, Friedman, and Durbin tests. Provided the mean of the ranks is $\frac{t^2+1}{2}$, this argument for the adjustment is for any treatment of ties, not just mid-ranks, and could also have been used for the CRD, RBD, and BIBD.

Given its derivation it would be natural to apply the RL test using the χ_{t-1}^2 sampling distribution, but simulations reported subsequently showed this was not appropriate. We were also concerned that although the parametric design is orthogonal the ranking necessary for the RL test could be considerably influenced by strong block effects. Aligning the raw data before ranking reduces this effect. The next section describes the alignment process.

9.4 Alignment

When ranking the data in the LSD ranking needs to be overall; ranking within blocks would be ambiguous and isn't an option. However, overall ranking may be compromised by a strong block effect.

Alignment (Hodges and Lehmann, 1962) is a technique often used in analysing data from factorial designs to obtain non-parametric tests of interaction effects that would otherwise be virtually inaccessible. The idea is to 'strip' away main effects by aligning and then ranking the aligned data. Resampling techniques can then be applied to these ranks. Use of alignment is appealing with Latin squares since it may be used to access the treatment effect after first stripping away the row and column effects. It is worth noting that the converse process, ranking first then aligning, gives the same ANOVA F test p-value as the unaligned ranked data. In other words, it is giving information about the ranked data, not the aligned and then ranked data.

We now show that alignment does eliminate the block effects.

Given the raw data $\{Y_{ijk}\}$ define $X_{ijk} = Y_{ijk} - \frac{Y_{\bullet j\bullet}}{t} - \frac{Y_{\bullet\bullet k}}{t} + \frac{Y_{\bullet\bullet\bullet}}{t^2}$; $\{X_{ijk}\}$ is the aligned data. We now show that when the parametric analysis is performed on the aligned data, the row and column sums of squares are zero, while the treatment and error sums of squares are as for the parametric analysis performed on the raw data. Thus, the treatments F statistic is the same for both raw and aligned data.

From the definition of X_{ijk}, it follows that

$$X_{\bullet\bullet\bullet} = Y_{\bullet\bullet\bullet} - Y_{\bullet\bullet\bullet} - Y_{\bullet\bullet\bullet} + Y_{\bullet\bullet\bullet} = 0,$$

$X_{i\bullet\bullet}$ (the sum over the t cells in which treatment i occurs) $= Y_{i\bullet\bullet} - \dfrac{Y_{\bullet\bullet\bullet}}{t}$

(each row is added once) $- \dfrac{Y_{\bullet\bullet\bullet}}{t}$ (each column is added once)

$+ \dfrac{Y_{\bullet\bullet\bullet}}{t} = Y_{i\bullet\bullet} - \dfrac{Y_{\bullet\bullet\bullet}}{t},$

$X_{\bullet j\bullet}$ (the sum over the t cells in row j) $= Y_{\bullet j\bullet} - Y_{\bullet j\bullet} - \dfrac{Y_{\bullet\bullet\bullet}}{t} + \dfrac{Y_{\bullet\bullet\bullet}}{t} = 0,$

$X_{\bullet\bullet k} = 0$ similarly.

Sums of squares for the aligned data will be given a suffix 'A'. From the previous text, it follows that

$$CFA = 0,$$

$$SSA_{\text{Treat}} = \sum_{i=1}^{t} \frac{X_{i\bullet\bullet}^2}{t} - CFA = \sum_{i=1}^{t} \frac{\left(Y_{i\bullet\bullet} - \frac{Y_{\bullet\bullet\bullet}}{t}\right)^2}{t} = \sum_{i=1}^{t} \frac{Y_{i\bullet\bullet}^2}{t} - CF = SS_{\text{Treat}},$$

$$SSA_{\text{Row}} = \sum_{j=1}^{t} \frac{X_{\bullet j\bullet}^2}{t} - CFA = 0, \text{ and } SSA_{\text{Col}} = 0 \text{ similarly.}$$

We now seek to prove that $SSA_{\text{Error}} = SS_{\text{Error}}$, or, equivalently, in light of the previous text,

$$SSA_{\text{Total}} = SS_{\text{Total}} - SS_{\text{Row}} - SS_{\text{Col}}.$$

Now

$$SSA_{\text{Total}} = \sum_{\text{all cells}} \left(Y_{ijk} - \frac{Y_{\bullet j \bullet}}{t} - \frac{Y_{\bullet \bullet k}}{t} + \frac{Y_{\bullet \bullet \bullet}}{t^2} \right)^2$$

$$= \sum_{\text{all cells}} \left[\left(Y_{ijk} - \frac{Y_{\bullet \bullet \bullet}}{t^2} \right) - \left(\frac{Y_{\bullet j \bullet}}{t} - \frac{Y_{\bullet \bullet \bullet}}{t^2} \right) - \left(\frac{Y_{\bullet \bullet k}}{t} - \frac{Y_{\bullet \bullet \bullet}}{t^2} \right) \right]^2$$

$$= \sum_{\text{all cells}} \left(Y_{ijk} - \frac{Y_{\bullet \bullet \bullet}}{t^2} \right)^2 + \sum_{\text{all cells}} \left(\frac{Y_{\bullet j \bullet}}{t} - \frac{Y_{\bullet \bullet \bullet}}{t^2} \right)^2$$

$$+ \sum_{\text{all cells}} \left(\frac{Y_{\bullet \bullet k}}{t} - \frac{Y_{\bullet \bullet \bullet}}{t^2} \right)^2 - 2 \sum_{\text{all cells}} \left(Y_{ijk} - \frac{Y_{\bullet \bullet \bullet}}{t^2} \right) \left(\frac{Y_{\bullet j \bullet}}{t} - \frac{Y_{\bullet \bullet \bullet}}{t^2} \right)$$

$$- 2 \sum_{\text{all cells}} \left(Y_{ijk} - \frac{Y_{\bullet \bullet \bullet}}{t^2} \right) \left(\frac{Y_{\bullet \bullet k}}{t} - \frac{Y_{\bullet \bullet \bullet}}{t^2} \right)$$

$$+ 2 \sum_{\text{all cells}} \left(\frac{Y_{\bullet j \bullet}}{t} - \frac{Y_{\bullet \bullet \bullet}}{t^2} \right) \left(\frac{Y_{\bullet \bullet k}}{t} - \frac{Y_{\bullet \bullet \bullet}}{t^2} \right).$$

We work through each expression in turn:

$$\sum_{\text{all cells}} \left(Y_{ijk} - \frac{Y_{\bullet \bullet \bullet}}{t^2} \right)^2 = SS_{\text{Total}},$$

$$\sum_{\text{all cells}} \left(\frac{Y_{\bullet j \bullet}}{t} - \frac{Y_{\bullet \bullet \bullet}}{t^2} \right)^2 = \sum_{j=1}^{t} Y_{\bullet j \bullet}^2 - CF = SS_{\text{Row}},$$

$$\sum_{\text{all cells}} \left(\frac{Y_{\bullet \bullet k}}{t} - \frac{Y_{\bullet \bullet \bullet}}{t^2} \right)^2 = SS_{\text{Col}},$$

$$\sum_{\text{all cells}} \left(Y_{ijk} - \frac{Y_{\bullet \bullet \bullet}}{t^2} \right) \left(\frac{Y_{\bullet j \bullet}}{t} - \frac{Y_{\bullet \bullet \bullet}}{t^2} \right) = \sum_{j=1}^{t} \left[\left(\frac{Y_{\bullet j \bullet}}{t} - \frac{Y_{\bullet \bullet \bullet}}{t^2} \right) \sum_{k=1}^{t} \left(Y_{ijk} - \frac{Y_{\bullet \bullet \bullet}}{t^2} \right) \right]$$

$$= \sum_{j=1}^{t} \left(\frac{Y_{\bullet j \bullet}}{t} - \frac{Y_{\bullet \bullet \bullet}}{t^2} \right)^2 = SS_{\text{Row}},$$

since $\sum_{k=1}^{t} \left(Y_{ijk} - \frac{Y_{\bullet \bullet \bullet}}{t^2} \right) = Y_{\bullet j \bullet} - \frac{Y_{\bullet \bullet \bullet}}{t}$. Continuing,

$$\sum_{\text{all cells}} \left(Y_{ijk} - \frac{Y_{\bullet \bullet \bullet}}{t^2} \right) \left(\frac{Y_{\bullet \bullet k}}{t} - \frac{Y_{\bullet \bullet \bullet}}{t^2} \right) = \sum_{k=1}^{t} \left[\left(\frac{Y_{\bullet \bullet k}}{t} - \frac{Y_{\bullet \bullet \bullet}}{t^2} \right) \sum_{k=1}^{t} \left(Y_{ijk} - \frac{Y_{\bullet \bullet \bullet}}{t^2} \right) \right]$$

$$= \sum_{\text{all cells}} \left(\frac{Y_{\bullet \bullet k}}{t} - \frac{Y_{\bullet \bullet \bullet}}{t^2} \right)^2 = SS_{\text{Col}},$$

similarly and

$$\sum_{\text{all cells}} \left(\frac{Y_{\bullet j \bullet}}{t} - \frac{Y_{\bullet \bullet \bullet}}{t^2} \right) \left(\frac{Y_{\bullet \bullet k}}{t} - \frac{Y_{\bullet \bullet \bullet}}{t^2} \right) = \sum_{j=1}^{t} \left(\frac{Y_{\bullet j \bullet}}{t} - \frac{Y_{\bullet \bullet \bullet}}{t^2} \right)$$

$$\times \sum_{k=1}^{t} \left(\frac{Y_{\bullet \bullet k}}{t} - \frac{Y_{\bullet \bullet \bullet}}{t^2} \right) = 0.$$

Thus, $SSA_{\text{Total}} = SS_{\text{Treat}} - SS_{\text{Row}} - SS_{\text{Col}}$ and hence, $SSA_{\text{Error}} = SS_{\text{Error}}$, as required.

Since $SSA_{\text{Treat}} = SS_{\text{Treat}}$ and $SSA_{\text{Error}} = SS_{\text{Error}}, AF = \frac{MSA_{\text{Treat}}}{MSA_{\text{Error}}} = \frac{MS_{\text{Treat}}}{MS_{\text{Error}}} = F$; the ANOVA F statistics for the raw and aligned data are equal. A corollary is that the row and column sums of squares for the aligned data are identically zero: $SSA_{\text{Row}} = SSA_{\text{Col}} = 0$. It is also trivially true that the grand total, and all row and column totals are zero.

Given that we know that alignment removes the row and column block effects, it may be tempting to apply a parametric analysis to the aligned data ignoring either or both rows and columns: effectively analysing the aligned data as completely randomised or randomised block data. The idea is to 'gift' block degrees of freedom to the error. However, such analyses using standard software won't recognise the linear constraints of the alignment process, from which it follows that the data are no longer mutually independent. Thus, if the one-factor F test statistic is used to test if the treatment means are equal, the F statistic does not follow an F distribution: such an analysis is invalid.

A parametric analysis of the aligned data results in the block sums of squares being zero and hence, the block p-values are precisely one. However, if the aligned data are ranked and then the parametric analysis applied, in all the data sets, we have analysed the block p-values are close to but less than one. It seems ranking after aligning doesn't quite annihilate the block effects.

Ranking is perhaps most frequently used when the original parametric analysis is dubious because the model assumptions are compromised. One obvious way forward would then be to use resampling methods to obtain p-values; after all, the F test statistic in the LSD is still a valid test statistic in that large values are consistent with rejection of the null hypothesis. If using resampling methods is not an option, then ranking is a sensible approach. Of course, it may be that the experimenter feels that analysis of the ranks is more meaningful than analysis of the raw data.

Any formal statement of the conclusion should indicate that a location treatment effect has been indicated – or not – by analysis of the ranked data or aligned then ranked data. We emphasise that the aligned tests require the raw data, since it is these that are aligned and then ranked.

9.5 Simulation Study

Before embarking on a simulation study, we need to explain why we have chosen to go about it as we have. We are aware of Akritas (1990) and others who built on that work. The idea is that if the parametric model was found to be valid, rank methods transform that model, and the consequences of this was explored for various rank transform (RT) scenarios. This is not the position we take.

As we noted earlier, rank methods are often used when the parametric model is found to be dubious. It is then reasonable to transform the data and consider inference on a new model for the transformed data. Even if the parametric model is not found to be dubious, it is not unreasonable to consider other models. After all there is a considerable literature on model selection, and it is not uncommon for multiple models to be consistent with the data. The question then is which is to be preferred. Hence we are prepared to accept inference based on the rank transform method if the model for the ranked data is valid. It perhaps needs to be underlined that such inference is based not on the rank transformed model but an independent parametric model for the ranks.

In the following simulation study, we wish to compare tests based on quite different models. To do so we assume a single model, the parametric model based on the raw data described in Section 9.2. We test that the vector of treatment effects, (γ_k), is zero against not this. This may be a harsh examination for the rank based tests that may perform quite differently for a more appropriate model. Nevertheless, the study produces very useful results.

Initial pilot simulation testing of the non-aligned *RL* statistic indicated that the use of the χ^2_{t-1} sampling distribution statistic worked quite well in the scenario where no block effects were present, with test sizes being close to the nominal significance level. However, for the LSD, this is not a realistic scenario, as one would expect at least one of the blocking variables to influence the response variable. When data are simulated with block effects, the response values are contaminated by the effect of the blocking variables. This results in a test with extremely small test sizes with an inability to reject the null hypothesis.

The aligned *RL* (*ARL*) statistic first aligns to remove the contamination of the block effects, and then ranks the data. This does not totally remove the block effects from this statistic, but as subsequent demonstrate, the block effects are considerably reduced. However, the aligned and then ranked data are no longer independent. A pilot assessment of the χ^2_{t-1} sampling distribution for *ARL* found that test sizes were too large. For a nominal significance level of 0.05, and for the variety of input parameters considered, the test sizes ranged roughly between 0.15 and 0.25. Exploratory analysis indicated that while the sampling distribution was not χ^2_{t-1}, it is similar to χ^2 with some other degrees of freedom.

As in previous chapters, here we also take the opportunity to assess the effect of categorisation on the competing tests. Of course, for categorised data, the test statistic is discrete. For heavy categorisation, the number of values a test statistic may take is relatively small. As the number of categories increases, so does the number of values the test statistic takes, and the probability function approaches that of the continuous test statistic: the no categorisation case.

A simulation study was formulated to generate sets of sampling distributions based on varying input parameters with the goal of equating the sampling distribution means with the mean of a χ^2_ν distribution. The input parameters for the study were the following: $t = 3, \ldots, 10$, row effect multipliers $d_{\text{row}} = -1, 0, 1$, column effect multipliers $d_{\text{col}} = -1, 0, 1$ and categorisation parameters $c = 0, 3, 10, 20$. The row effects are set by creating a sequence from -3 to 3 with length equal to t and then multiplying by the row multiplier d_{row}. Similarly, the column effects are set by taking a sequence from -2 to 2 with length t and then multiplying it by the column effect multiplier d_{col}. To generate data, first errors are randomly generated on the interval $(-1, 1)$ using a beta distribution, to be described shortly. These errors are then used with the relevant row and column effects to create a continuous response variable. The input parameter c defines the level of categorisation of the response variable. To achieve this, the range of the response variable is sliced into c intervals. If a continuous response falls in the ith interval, the categorised response variable then takes the value i. A value of $c = 0$ represents no categorisation, whereas values of $c = 3, 10$, and 20 represent relatively high, medium, and low levels of categorisation, respectively.

For each set of inputs, 100,000 data sets are generated, and the test statistic calculated. Subsequently, the mean of those test statistics is calculated. This is then repeated 30 times with a new random LSD simulated each time. Thus, a total of 4320 distribution means are calculated for the *ARL* test statistic.

Figure 9.1 shows a plot of the means of the empirical sampling distributions against the number of treatments t. The oblique lines in the left panels indicate the line $y = t + 1$. When there is no categorisation of

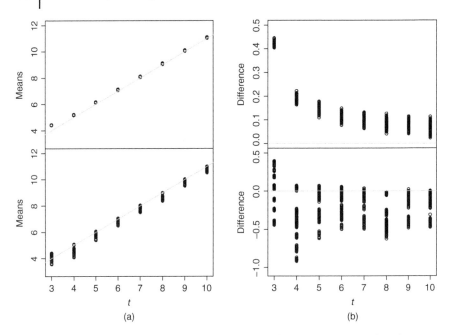

Figure 9.1 A scatter plot for the means of the ARL test statistics against the number of treatments t (a). Scatter plots of the difference between the means of the ARL test statistics and the line $y = t + 1$ (b). The top images are when there is no categorisation of the response variable and the bottom plots are for when $c = 10$.

the response variable, this appears to slightly underestimate the points for lower treatments sizes. The lower plots are for when $c = 10$. There it appears that $t + 1$ slightly overestimates the means on average. With these results in mind, it was decided that as a reasonable compromise for the tables likely to be used in practice, $v = t + 1$ be used for the degrees of freedom in a modified version of the ARL test designated as the CARL test. In the event a practitioner was uncomfortable using the estimated degrees of freedom in the CARL test, they may wish to do a permutation test based on the ARL test. We refer to this test as the PARL.

To calculate a permutation test p-value such as for the PARL, the raw data are aligned and ranked, and then RL is calculated, giving the observed test statistic, ARL_{obs}, say. Then for each permutation, the aligned data are permuted and the block effects, calculated from the raw data, are added back. This gives the permuted data, which are then aligned and RL calculated. This is repeated for the number of permutations considered with the p-value calculated as the proportion of these RL greater than ARL_{obs}.

A second simulation study was employed to assess the test size and power of several tests that are applicable to the LSD. The tests used are the following:

- the parametric F test,
- the F test applied to the ranks, the rank transform (RT) test,
- the F test applied to the aligned rank transform (FART),
- the CARL and,
- the PARL.

The input parameters for this simulation study are the following: $t = 4, 5$, and 6, multipliers $d_{tre} = 0.00, 0.03, 0.06, \ldots, 1.50$, $d_{row} = 0, 1$, $d_{col} = -1, 1$, and $c = 0, 3, 10, 20$. This is repeated for three different error distributions that we denote as $D = 1, 2, 3$. The treatment effects assess realistic LSDs used in practice. The d_{tre} multiplier is used similarly to d_{row} and d_{col} as described earlier. It is a multiplier for a vector of treatments effects. The vectors of row and column effects are no longer calculated as a linear interpolation between two constants as done in the previous study. For each value of t, the row and column effects together with the treatment effects are shown in Table 9.1.

A value of $d_{tre} = 0$ represents a calculation of the test size and $d_{tre} > 0$ is an assessment of power. The first error distributions is a beta distribution with parameters are $\alpha_\epsilon = \beta_\epsilon = 0.5$. This corresponds to a U-shaped distribution. The second error distribution is again a beta distribution with parameters $\alpha_\epsilon = 5, \beta_\epsilon = 1$. This is a skewed distribution. Both of these distributions were re-scaled into an interval of length 2, and shifted such that the population mean of the distribution is located on 0. The third error distribution is a mixture of a gamma distribution and a normal distribution. This results in a distribution skewed to the right and allows for outliers in the right tail. The input parameter c is used as described earlier.

For each set of parameters, 10,000 random data sets are generated and the number of rejections for each test are recorded. For the PARL, 2500 permutations of the data set are performed.

Table 9.1 Row and column block effects together with treatment effects for each value of t.

Effect	$t = 4$	$t = 5$	$t = 6$
Row	$(-4, -1, 2, 3)$	$(-3, -0, 1, 1, 1)$	$(-2, -1, 0, 0, 1, 2)$
Column	$(-3, 0, 1, 2)$	$(-2, -1, 0, 1, 2)$	$(-2, -1, -1, 0, 1, 3)$
Treatment	$(-3, -1, 2, 2)$	$(-2, -1, 0, 0, 3)$	$(-3, -1, -0, 1, 1, 2)$

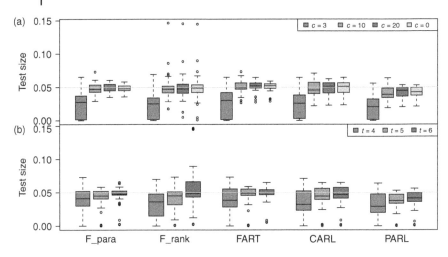

Figure 9.2 Side-by-side boxplots showing test sizes for each of the tests broken down by categorisation (a) and number of treatments (b).

Figure 9.2 shows the test sizes for each of the tests separated by the categorisation input parameter c and the number of treatments input parameter t, for all three error distributions. Overall, the tests do reasonably well in adhering to the nominal significance level. There are some input parameters that appear to have a greater effect than others. The categorisation parameter caused the most variation in test sizes followed by the number of treatments parameter. For a high level of categorisation ($c = 3$), there was a greater variation in test sizes, and on average, these were well below the nominal significance level. For the number of treatments, with $t = 4$, there was greater variation than in larger values of t. For $t = 5, 6$ the test sizes improved and showed less variation for all the tests except the F test on the ranks.

Figure 9.3 shows power curves for a very small subset of input parameters. Of course, the parameter spaces are multi-dimensional and our treatment allows one-dimensional slices of the power curves. The left panels give the powers, while the right panels give differences between powers for each test with the parametric ANOVA, the gold standard. Positive differences indicate better performance than the parametric ANOVA.

The first thing to note is that all the tests appear to perform quite similarly except for the F test on the ranks, the rank transform test. It sometimes performs adequately, but at other times, the performance is poor. Where no categorisation of the response variable is employed, the tests have greatest power, needing a relatively small value of d_{tre} to reach 100%

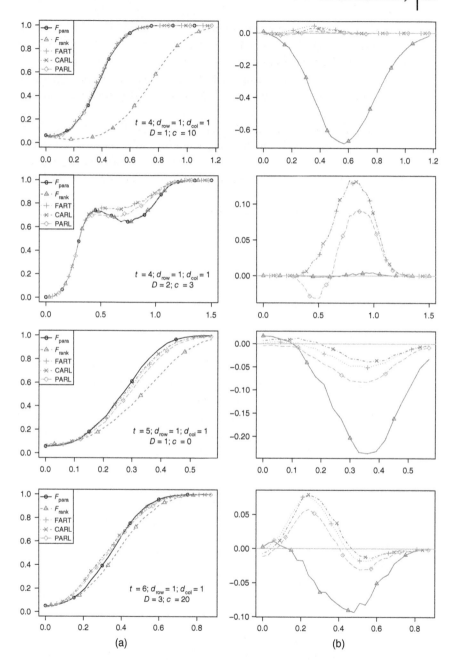

Figure 9.3 A subset of indicative power curves generated for the indicated input parameter values (a) together with plots of the difference in power between each test and the parametric ANOVA (b).

power. As categorisation moves from high to low, power tends to increase for a given value of d_{tre}. For high categorisation and a U-shaped error distribution, the non-parametric FART and CARL appear to outperform the parametric ANOVA. Surprisingly, the CARL tends to always outperform the PARL in terms of power. The FART and CARL results are always very similar.

Overall, the parametric ANOVA, FART, and CARL perform quite well across the range of input parameters. The PARL also performs satisfactorily. However, given the additional effort required to perform such a test and its performance relative to the CARL, one would likely tend to use the latter. For some input parameters, the F test on the ranks performs quite poorly relative to the other tests and therefore should be avoided.

9.6 Examples

9.6.1 Dynamite Data

An experimenter is studying the effect of five different formulations of an explosive mixture. A batch of raw material is large enough for only five formulations. Each formulation is prepared by five operators. The response here is the explosiveness of dynamite formulations. The data are in Table 9.2. The layout is a standard Latin square with the first row in alphabetical order. At the time of writing, the dynamite scenario and data could be accessed from https://www.fox.temple.edu/cms/wp-content/uploads/2016/05/Randomized-Block-Design.pdf.

Table 9.2 Dynamite data cross-tabulated with batches and operators, with treatments denoted by letters A–E together with the raw response values in each cell.

	Operators				
Batches	**1**	**2**	**3**	**4**	**5**
1	A 24 (11.5)	B 20 (4.5)	C 19 (3)	D 24 (11.5)	E 24 (11.5)
2	B 17 (1)	C 24 (11.5)	D 30 (20.5)	E 27 (17.5)	A 36 (24)
3	C 18 (2)	D 38 (25)	E 26 (15)	A 27 (17.5)	B 21 (6)
4	D 26 (15)	E 31 (22.5)	A 26 (15)	B 23 (9)	C 22 (7.5)
5	E 22 (7.5)	A 30 (20.5)	B 20 (4.5)	C 29 (19)	D 31 (22.5)

Mid-ranks are in parentheses.

The parametric analysis gives an F value of 7.7344 with an $F_{4,12}$ p-value 0.0025. The Shapiro–Wilk test for normality of the residuals has a p-value of 0.0581, indicating some doubt about the validity of the parametric analysis and suggesting a non-parametric approach would be less problematic. The parametric analysis of the overall mid-ranks, the rank transform test, gives an $F_{4,12}$ p-value 0.0002. These residuals are consistent with normality.

To apply the tests based on RL, we need the treatment rank sums, which, for the non-aligned test, are 88.5, 25, 43, 94.5, and 74 for treatments for A–E, respectively. The corresponding rank sums for the aligned test are 90, 21, 42, 103.5, and 68.5.

The p-values for the FART and CARL are 0.0027 and 0.0071, and, using 100,000 permutations, 0.0042 for the PARL. All are consistent with 0.0025 for the F test on the raw data and 0.0002 for the rank transform test. There is clear evidence of differences in the treatment means.

```
# F_para
Anova(lm(response ~ treatment + batch + operator),type = 3)[2,]

## Anova Table (Type III tests)
##
## Response: response
##              Sum Sq Df F value    Pr(>F)
## treatment      330  4  7.7344 0.002537 **
## ---
## Signif. codes:  0 '***' 0.001 '**' 0.01 '*' 0.05 '.' 0.1 ' ' 1

# F_rank
Anova(lm(rank(response)~treatment+batch+operator),type = 3)[2,]

## Anova Table (Type III tests)
##
## Response: rank(response)
##              Sum Sq Df F value     Pr(>F)
## treatment     717.5  4  13.093 0.0002468 ***
## ---
## Signif. codes:  0 '***' 0.001 '**' 0.01 '*' 0.05 '.' 0.1 ' ' 1

# FART
row_col_eff = row_col_effect(response, operator, batch)
A_resp = response - row_col_eff
Anova(lm(rank(A_resp) ~ treatment + batch + operator),type = 3)[2,]

## Anova Table (Type III tests)
##
## Response: rank(A_resp)
##              Sum Sq Df F value    Pr(>F)
```

```
## treatment  916.9  4  7.5944 0.002733 **
## ---
## Signif. codes:  0 '***' 0.001 '**' 0.01 '*' 0.05 '.' 0.1 ' ' 1

# CARL
CARL(y = response, treatment = treatment, block1 = batch,
     block2 = operator)

##
##        CARL Test for the Latin Square Design
##
## data:  response, treatment,
##        batch and operator
##
## Test statistic = 17.68029, df = 6, p-value = 0.007082857
##
## For the group levels 1 2 3 4 5,
## the aligned rank sums are:  90 21 42 103.5 68.5

# PARL
PARL(y = response, treatment = treatment, block1 = batch,
     block2 = operator, N_perms = 100000)

##
##        PARL Test for the Latin Square Design
##             with 1e+05 permutations
##
## data:  response, treatment,
##        batch and operator
##
## Test statistic = 17.68029, p-value = 0.00415
##
## For the group levels 1 2 3 4 5,
## the aligned rank sums are:  90 21 42 103.5 68.5
```

9.6.2 Peanuts Data

A plant biologist conducted an experiment to compare the yields of four varieties of peanuts, denoted as A, B, C, and D. A plot of land was divided into 16 subplots (four rows and four columns). The LSD implicit in Table 9.3 was run. The responses are also given in Table 9.3. At the time of writing the peanut varieties, data were available at http://www.math.montana.edu/jobo/st541/sec3c.pdf.

In the parametric analysis, peanuts have a p-value of 0.0870, while the Shapiro–Wilk test for normality of the residuals has a p-value of 0.0725. If the same analysis is performed on the data ranked overall, the peanut p-value is 0.0889, while the residuals are consistent with normality.

Table 9.3 Peanut yields and the corresponding varieties.

	Columns			
Row	**1**	**2**	**3**	**4**
1	26.7 (C)	19.7 (A)	29.0 (B)	29.8 (D)
2	23.1 (A)	21.7 (B)	24.9 (D)	29.0 (C)
3	29.3 (B)	20.1 (D)	29.0 (C)	27.3 (A)
4	25.1 (D)	17.4 (C)	28.7 (A)	35.1 (B)

The RL tests require the mid-rank sums. For the non-aligned data for treatments A–D, these are 26, 46, 33, and 31, respectively, while for the aligned data, they are 22, 58, 34, and 22, respectively.

The *p*-values for the FART and CARL are 0.0763 and 0.0707, and, using 100,000 permutations 0.0881 for the PARL. At the 0.05 level, there is no evidence of a difference in the treatment means.

```
# F_para
Anova(lm(yield ~ treatment + row + col), type = 3)[2,]

## Anova Table (Type III tests)
##
## Response: yield
##            Sum Sq Df F value Pr(>F)
## treatment 42.667  3   3.558  0.087.
## ---
## Signif. codes:  0 '***' 0.001 '**' 0.01 '*' 0.05 '.' 0.1 ' ' 1

# F_rank
Anova(lm(rank(yield) ~ treatment + row + col), type = 3)[2,]

## Anova Table (Type III tests)
##
## Response: rank(yield)
##            Sum Sq Df F value  Pr(>F)
## treatment    54.5  3  3.5161 0.08887.
## ---
## Signif. codes:  0 '***' 0.001 '**' 0.01 '*' 0.05 '.' 0.1 ' ' 1

# FART
row_col_eff = row_col_effect(yield, row, col)
A_resp = yield - row_col_eff
Anova(lm(rank(A_resp) ~ treatment + row + col), type = 3)[2,]

## Anova Table (Type III tests)
##
```

```
## Response: rank(A_resp)
##            Sum Sq Df F value  Pr(>F)
## treatment    216  3   3.823 0.07629 .
## ---
## Signif. codes:  0 '***' 0.001 '**' 0.01 '*' 0.05 '.' 0.1 ' ' 1

# CARL
CARL(y = yield, treatment = treatment,
     block1 = row, block2 = col)

##
##         CARL Test for the Latin Square Design
##
## data:  yield, treatment,
##        row and col
##
## Test statistic = 10.16471, df = 5, p-value = 0.07070065
##
## For the group levels 1 2 3 4,
## the aligned rank sums are:  22 58 34 22

# PARL
PARL(y = yield, treatment = treatment,
     block1 = row, block2 = col, N_perms = 100000)

##
##         PARL Test for the Latin Square Design
##                with 1e+05 permutations
##
## data:  yield, treatment,
##        row and col
##
## Test statistic = 10.16471, p-value = 0.08814
##
## For the group levels 1 2 3 4,
## the aligned rank sums are:  22 58 34 22
```

9.6.3 Traffic Data

These data were considered in Section 8.7, where it was found that the NP ANOVA on the overall ranks gave treatment p-values of 0.0328, 0.9950, and 0.3823 for first, second, and third degrees, respectively. A similar analysis on the aligned overall ranks gave treatment p-values of 0.0743, 0.1655, and 0.2240. Neither analysis found higher-degree effects, and the aligned analysis was less critical of the null hypothesis than the unaligned analysis.

The non-aligned rank sums are 72, 74, 47, 62, and 70 for A–E, respectively, while if the data aligned before ranking the corresponding rank sums are 76, 94, 22, 54, and 79 for A–E, respectively.

The *p*-values for the FART and CARL are 0.0743 and 0.0613, and, using 100,000 permutations 0.0771 for the PARL. At the 0.05 level, there is no evidence of a difference in the treatment means. The non-significance of the aligned tests suggests the block effects have induced a spurious treatment effect.

```
# FART
row_col_eff = row_col_effect(minutes, intersection, time_of_day)
A_resp = minutes - row_col_eff
Anova(lm(rank(A_resp) ~ treatment + time_of_day + intersection),
      type = 3)[2,]

## Anova Table (Type III tests)
##
## Response: rank(A_resp)
##           Sum Sq Df F value  Pr(>F)
## treatment  625.6  4  2.8045 0.07431.
## ---
## Signif. codes:  0 '***' 0.001 '**' 0.01 '*' 0.05 '.' 0.1 ' ' 1

# CARL
CARL(y = minutes, treatment = treatment,
     block1 = intersection, block2 = time_of_day)

##
##         CARL Test for the Latin Square Design
##
## data:  minutes, treatment,
##        intersection and time_of_day
##
## Test statistic = 12.03077, df = 6, p-value = 0.06128589
##
## For the group levels 1 2 3 4 5,
## the aligned rank sums are:  76 94 22 54 79

# PARL
PARL(y = minutes, treatment = treatment,
     block1 = intersection, block2 = time_of_day, N_perms = 100000)

##
##         PARL Test for the Latin Square Design
##                with 1e+05 permutations
##
## data:  minutes, treatment,
##        intersection and time_of_day
##
## Test statistic = 12.03077, p-value = 0.07712
##
## For the group levels 1 2 3 4 5,
## the aligned rank sums are:  76 94 22 54 79
```

9.7 Orthogonal Trend Contrasts for Ordered Treatments

In Section 5.5, we constructed two sets of orthogonal trend contrasts for balanced designs when the treatments are ordered. The LSD is a balanced design so those constructions routinely apply here. We now find orthogonal contrasts for the examples examined in the previous section by decomposing the RL statistic. Although the RL test has been validated as a test of the treatment mean responses, it is also a test of equality of treatment rank means, as is shown in Section 9.8. The orthogonal decompositions give information about how those rank means may differ, whether or not the RL test suggests there is a difference.

With the definitions in Section 5.5, put $Y_i = \left(R_{i\bullet\bullet} - t\mu\right)\sqrt{d}$ in which $d = \left[\left(\sum_{j=1}^{t}\sum_{k=1}^{t}\frac{r_{jk}^2}{t}\right) - \frac{t(t^2+1)^2}{4}\right]^{-1}$ so that $RL = \sum_{i=1}^{t} Y_i^2$. Then $1^{T}Y = 0$ and the $c_s^{T}Y$ are orthogonal χ^2 contrasts of RL. The treatment sum of squares is $\sum_{i=1}^{t}\frac{(R_{i\bullet\bullet}-t\mu)^2}{t}$, so we take $Z_i = \frac{R_{i\bullet\bullet}-t\mu}{\sqrt{t}}$. The $m_s^{T}Z$ are used to construct orthogonal F contrasts.

We investigated approximating the sampling distributions of the RL χ^2 and F contrasts and found nothing that was satisfactory. We concluded that it was best to use permutation testing to find contrast p-values. In revisiting the LSD examples considered so far, we therefore focus on the FART and PARL tests and their components.

9.7.1 Dynamite Data Revisited

We previously found p-values for the FART and CARL are 0.0027 and 0.0071, and, using 100,000 permutations, 0.0039 for the PARL.

Using 100,000 permutations p-values for the aligned rank transform statistic and its linear, quadratic and cubic F contrasts were found to be 0.0039, 0.3252, 0.0384, and 0.0007, respectively.

Based on 100,000 permutations, the p-values for ARL and its linear, quadratic and cubic χ^2 contrasts are now 0.0037, 0.5433, 0.1477, and 0.0010, respectively.

A plot of the means of the aligned and then ranked data against dynamites given in Figure 9.4 suggests a cubic effect: that the rank means decrease, increase, and then decrease again. The contrast analysis supports this with a statistically significant cubic effect. This suggests differences are due primarily to third and fourth degree effects.

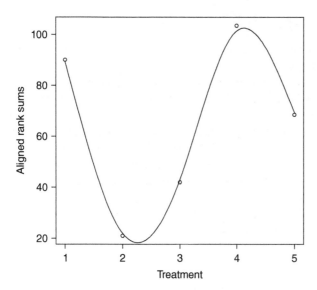

Figure 9.4 Rank sums of the aligned and then ranked dynamite data plotted against dynamites.

```
# PARL
PARL(y = response, treatment = treatment, block1 = batch,
     block2 = operator, N_perms = 100000, components = T)

##
##          PARL Test for the Latin Square Design
##                 with 1e+05 permutations
##
## data:   response, treatment,
##         batch and operator
##
## Test statistic = 17.68029, p-value = 0.00371
##
## For the group levels 1 2 3 4 5,
## the aligned rank sums are:   90 21 42 103.5 68.5
##
## Chi-squared components and F components:
##           Chi2 Obs stats p-value F Obs stats p-value
## Overall        17.6800 0.00371      7.5940 0.00389
## Degree 1        0.6017 0.54330      1.0340 0.32520
## Degree 2        3.2430 0.14770      5.5720 0.03842
## Degree 3       13.4100 0.00098     23.0500 0.00068
## Degree 4        0.4218 0.61150      0.7247 0.40940
```

9.7.2 Peanut Data Revisited

We previously found *p*-values for the FART, CARL, and PARL were all between 0.05 and 0.10.

With 100,000 permutations the *p*-values for the aligned RL statistic and its linear, quadratic, and cubic χ^2 contrasts are 0.0896, 0.6885, 0.0515, and 0.2032, respectively.

Using 100,000 permutations *p*-values for *F* and its linear, quadratic, and cubic contrasts were found to be 0.0871, 0.5518, 0.0371, and 0.1188.

The differences in the mean ranks of the aligned data appear to be due primarily to a quadratic effect: the B and C varieties have greater yields than the A and D varieties.

In the more omnibus FART, CARL, and PARL, the other contrasts, most particularly the linear component, have masked a significant quadratic effect that is apparent in Figure 9.5.

```
# PARL
PARL(y = yield, treatment = treatment,
     block1 = row, block2 = col,
     N_perms = 100000, components = T)

##
##         PARL Test for the Latin Square Design
##                with 1e+05 permutations
##
## data:  yield, treatment,
##        row and col
##
## Test statistic = 10.16471, p-value = 0.08961
##
## For the group levels 1 2 3 4,
## the aligned rank sums are:  22 58 34 22
##
## Chi-squared components and F components:
##         Chi2 Obs stats p-value F Obs stats p-value
## Overall       10.1600 0.08961     3.8230 0.08710
## Degree 1       0.3388 0.68850     0.3823 0.55180
## Degree 2       6.7760 0.05150     7.6460 0.03711
## Degree 3       3.0490 0.20320     3.4410 0.11880
```

9.7.3 Traffic Data Revisited

Using 100,000 permutations, *p*-values for the *F* statistic and its linear, quadratic, cubic, and quartic contrasts were found to be 0.0811, 0.5276, 0.0830, 0.1424, and 0.0518, respectively. Using 100,000 permutations, the *p*-values for the aligned RL test and its linear, quadratic, cubic, and quartic

Figure 9.5 Rank sums of the aligned and then ranked peanuts data plotted against peanut varieties.

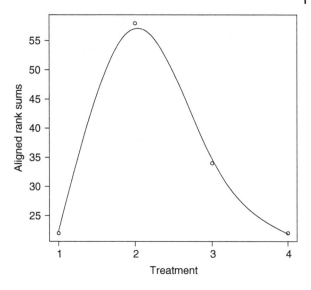

Figure 9.6 Rank sums of the aligned and then ranked traffic data plotted against traffic light signal sequences.

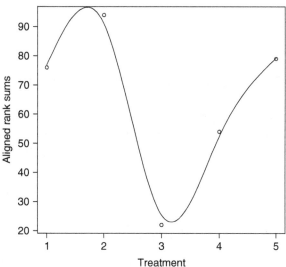

χ^2 contrasts were 0.0764, 0.6069, 0.1150, 0.1938, and 0.0680, respectively. Both the PARL and the FART find the treatment effect is not significant at the 0.05 level. There is almost a statistically significant quartic effect at the 0.05 level using both tests. The contrast p-values together with Figure 9.6 suggests that the relationship is a complex one. The masking of the higher-order effects is a persuasive reason for looking at component contrasts.

```
# PARL
PARL(y = minutes, treatment = treatment,
     block1 = intersection, block2 = time_of_day,
     N_perms = 100000, components = T)

##
##             PARL Test for the Latin Square Design
##                    with 1e+05 permutations
##
## data:   minutes, treatment,
##         intersection and time_of_day
##
## Test statistic = 12.03077, p-value = 0.07637
##
## For the group levels 1 2 3 4 5,
## the aligned rank sums are:   76 94 22 54 79
##
## Chi-squared components and F components:
##           Chi2 Obs stats p-value F Obs stats p-value
## Overall         12.0300 0.07637     2.8050 0.08110
## Degree 1         0.4446 0.60690     0.4146 0.52760
## Degree 2         3.8250 0.11500     3.5670 0.08297
## Degree 3         2.6500 0.19380     2.4710 0.14240
## Degree 4         5.1110 0.06803     4.7660 0.05182
```

9.8 Technical Derivation of the RL Test

Rayner and Beh (2009) give general results that include partitioning the Pearson statistic X_P^2 given a singly ordered four-way table of counts $\{N_{rijk}\}$ in which $r = 1, \ldots, n = t^2$, and the responses r are ordered while the independent variables, i, j, k, are not. For this table, if $p_{\bullet i \bullet \bullet} = \frac{n_{\bullet i \bullet \bullet}}{n}$, $p_{\bullet \bullet j \bullet} = \frac{n_{\bullet \bullet j \bullet}}{n}$, and $p_{\bullet \bullet \bullet k} = \frac{n_{\bullet \bullet \bullet k}}{n}$, define the components Z_{uijk} by

$$Z_{uijk} = \sqrt{\frac{n}{p_{\bullet i \bullet \bullet} p_{\bullet \bullet j \bullet} p_{\bullet \bullet \bullet k}}} \sum_{r=1}^{t^2} a_u(r) p_{rijk} = \sqrt{\frac{n^2}{n_{\bullet i \bullet \bullet} n_{\bullet \bullet j \bullet} n_{\bullet \bullet \bullet k}}} \sum_{r=1}^{t^2} a_u(r) n_{rijk}.$$

Rayner and Beh (2009) show that X_P^2 may be partitioned into components Z_{uijk} by $X_P^2 = \sum_{u=1}^{n-1} \sum_{i=1}^{I} \sum_{j=1}^{J} \sum_{k=1}^{K} Z_{uijk}^2$.

Now assume the smooth model

$$p_{rijk} = p_{r\bullet\bullet\bullet} \left[1 + \sum_{u=1}^{n-1} \theta_{uijk} a_u(r) \right],$$

in which the θ_{uijk} are real valued parameters and the $\{a_u(r)\}$ are orthonormal on $\{p_{r\bullet\bullet\bullet}\}$. In Rayner and Best (2013, Appendix B), it is shown that $E[Z_{uijk}] = \theta_{rijk}$ and that under the null hypothesis that all the θ_{uijk} are zero, or assuming that the θ_{rijk} are $O\left(n^{-0.5}\right)$, all $Var\left(Z_{uijk}\right)$ are or tend to one and, for $(u,i,j,k) \neq (u',i',j',k')$, all $Cov\left(Z_{uijk}, Z_{u'i'j'k'}\right)$ tend to zero. Since by the CLT the Z_{uijk} approach normality, they are all asymptotically mutually independent and asymptotically standard normal. The same is true of the $Z_{ui\bullet\bullet}$.

For the LSD $n_{\bullet i\bullet\bullet} = n_{\bullet\bullet j\bullet} = n_{\bullet\bullet\bullet k} = t$ and $n = t^2$. If ranking is overall, from 1 to t^2, then $Z_{uijk} = \sqrt{t}\sum_{r=1}^{t^2} a_u(r)n_{rijk}$. Suppose we apply the Latin square ANOVA to the $\left\{Z_{uijk}\right\}$ for a given $u \geq 1$. Given the conclusions about the distribution of the Z_{uijk} in the previous paragraph, this is asymptotically valid. Whether or not there are ties, for any observation in such a Latin square, there is only one value r for which n_{rijk} is non-zero, and then it is one. This corresponds to precisely one (i,j,k) triple. Thus, for a given u, for each observation, there exist an r and a triple (i,j,k) for which $z_{uijk} = \sqrt{t}\,a_u(r)$. Hence, without loss of generality, the Latin square ANOVA can be applied to the $\left\{a_u(r)\right\}$ for each u. This independently justifies the nonparametric ANOVA analysis validated in the Chapter 8 for designs consistent with the general linear model.

Now

$$Z_{u\bullet\bullet\bullet} = \sqrt{t}\sum_{r=1}^{t^2}\sum_{i=1}^{I}\sum_{j=1}^{J}\sum_{k=1}^{K} a_u(r)n_{rijk} = \sqrt{t}\sum_{r=1}^{t^2} a_u(r)n_{r\bullet\bullet\bullet}$$

$$= \sqrt{tn}\sum_{r=1}^{t^2} a_u(r)p_{r\bullet\bullet\bullet} = 0,$$

using the orthogonality of $\left\{a_u(r)\right\}$. Since $Z_{u\bullet\bullet\bullet} = 0$, using the results from Section 9.2, $CF = 0$, the treatment sum of squares is $\sum_{i=1}^{I}\frac{Z_{ui\bullet\bullet}^2}{t}$, the row sum of squares is $\sum_{j=1}^{J}\frac{Z_{u\bullet j\bullet}^2}{t}$, the column sum of squares is $\sum_{k=1}^{K}\frac{Z_{u\bullet\bullet k}^2}{t}$, the total sum of squares is $\sum_{i=1}^{I}\sum_{j=1}^{J}\sum_{k=1}^{K} Z_{uijk}^2 = t\sum_{r=1}^{t^2} a_u^2(r)$ and if there are no ties, the latter is $nt\sum_{r=1}^{t^2} a_u^2(r)p_{r\bullet\bullet\bullet} = nt$. The error sum of squares is found by difference.

Moreover, since the Z_{uijk} are all asymptotically mutually independent and asymptotically normal, the $Z_{ui\bullet\bullet}$ are all asymptotically mutually independent and asymptotically normal with variance t^2. Now, we know that if $X_1,\ldots,X_n$ are mutually independent $N(\mu,\sigma^2)$ random variables, then $\sum_{i=1}^{I}\frac{\left(X_i-\bar{X}\right)^2}{\sigma^2}$ is χ_{n-1}^2 distributed. Thus, for each u under the null hypothesis that all $E\left[Z_{ui\bullet\bullet}\right]$ are zero $\sum_{i=1}^{I}\frac{Z_{ui\bullet\bullet}^2}{t}$ is asymptotically χ_{t-1}^2 distributed.

Consider now the case of degree one: $u = 1$. If there are no ties, the ranks have mean $\mu = \frac{t^2+1}{2}$ and variance $\sigma^2 = \frac{(t^2-1)(t^2+1)}{12}$. If there are ties, then σ^2

may be modified as previously. With these values

$$Z_{1ijk} = \sqrt{t} \sum_{r=1}^{t^2} \frac{r - \mu}{\sigma} n_{rijk} = t^{0.5} \frac{R_{ijk} - \mu}{\sigma},$$

since $n_{\bullet ijk} = 1$ (a particular treatment must be assigned some rank). Here R_{ijk} is the rank of treatment i in row j and column k. Thus, the $\frac{Z_{1i\bullet\bullet}}{\sqrt{t}} = \frac{R_{i\bullet\bullet} - t\mu}{\sigma}$ have sample means zero and are asymptotically normal with variance t. Thus, under the null hypothesis that all $E\left[Z_{1i\bullet\bullet}\right]$ are zero, $\sum_{i=1}^{I} \frac{(R_{i\bullet\bullet} - t\mu)^2}{t\sigma^2}$ is asymptotically χ_{t-1}^2 distributed. This null hypothesis is equivalent to all rank sums having the same expectation. Thus, the RL test is a test of equality of treatment mean ranks. We have established empirically that it is also a test of equality of treatment means.

Bibliography

Akritas, M. G. (1990). The rank transform method in some two-factor designs. *Journal of the American Statistical Association*, 85(409):73–78.

Hodges, J. L. and Lehmann, E. L. (1962). Rank methods for combination of independent experiments in analysis of variance. *The Annals of Mathematical Statistics*, 33(2):482–497.

Rayner, J. C. W. and Beh, E. J. (2009). Components of Pearson's statistic for at least partially ordered m-way contingency tables. *Advances in Decision Sciences*, 2009: Article ID 980706 9, https://doi.org/10.1155/2009/980706, 2009.

Rayner, J. C. W. and Best, D. J. (2013). Extended ANOVA and rank transform procedures. *Australian & New Zealand Journal of Statistics*, 55(3):305–319.

10

Ordered Non-parametric ANOVA

10.1 Introduction

Everything said in the Introduction to Chapter 8 is relevant here – but now we focus on scenarios in which the treatments are ordered. In Section 8.1, there is a description of ordered NP ANOVA, and perhaps a reasonable way to reinforce that is to apply the approach to the Strawberry data.

10.1.1 Strawberry Data Revisited

We found in Section 8.1, that as well as the strong treatment effect found by Pearce, there was a significant second-degree effect that appears to reflect differences in pesticide variances.

We now assume there is a natural ordering of the pesticides: A, B, C, D, O. A plot of response versus pesticide, along with two smooth lines fitted to the data, is given in Figure 10.1. The plot suggests that although there is a tendency for the response to increase as we go from pesticide A to pesticide O, the relationship may be more complex than that.

To investigate this, we examine the generalised correlations for these data. Generalised correlations were discussed and described in Section 6.3. There we defined the (u, v)th generalised correlation test statistic to be

$$V_{uv} = \sum_{i=1}^{I} \sum_{j=1}^{J} \frac{N_{ij} a_u (x_i) b_v (y_j)}{\sqrt{n}}.$$

We test the nine null hypotheses that the generalised correlations for u and $v = 1, 2, 3$ are in each case zero against the alternative that it is non-zero. See Table 10.1. In each cell, there are two numbers. The first is the one sample t-test p-value when testing for consistency with zero, while the second is the corresponding Wilcoxon signed rank test p-value. The Shapiro–Wilk normality test finds the residuals of the various ANOVAs

An Introduction to Cochran–Mantel–Haenszel Testing and Nonparametric ANOVA, First Edition. J.C.W. Rayner and G. C. Livingston Jr.
© 2023 John Wiley & Sons Ltd. Published 2023 by John Wiley & Sons Ltd.

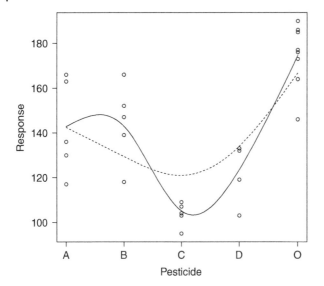

Figure 10.1 Scatter plot of the response variable versus pesticide type for the strawberry data together with two smooth splines with different smoothing parameters.

Table 10.1 *P*-values for generalised correlations tested for consistency with zero.

Response order	Pesticide order		
	1	**2**	**3**
1	0.0467, 0.0587	0.0004, 0.0006	0.0078, 0.0072
2	0.0209, 0.0256	0.5587, 0.6004	0.6819, 0.3055
3	0.5618, 0.6987	0.8118, 0.9093	0.1929, 0.1580

In each cell, the first entry is the one sample *t*-test *p*-value, while the second is the Wilcoxon signed rank test *p*-value.

are all consistent with normality except that for the (2, 3)th generalised correlation. This correlation is not significant by either the t or signed rank test.

The (1, 1), (1, 2), (1, 3), and (2, 1) cells in Table 10.1 have t-test p-values of less than 0.05. The significance of these generalised correlations reflects the fact that the mean responses need at least a cubic relation to model; reflected by the smooth line in Figure 10.1. See Section 10.6.

Table 10.2 ANOVA *F* test *p*-values assessing if generalised correlations vary across levels of the blocks.

Response order	Product order		
	1	2	3
1	0.4232	0.6568	0.7002
2	0.4470	0.4771	0.5997
3	0.0663	0.4249	0.3038

The significant (2, 1) generalised correlation suggests an umbrella effect: as we go from pesticide A to O, there is a quadratic effect in the responses, an umbrella effect. This is somewhat apparent in Figure 10.1 looking at the smooth fit with greater smoothing (dashed line).

The ordered NP ANOVA allows us to assess whether or not the generalised correlations vary across the levels of the independent variables. See Table 10.2. In all cases, the Shapiro–Wilk tests find the residuals are consistent with normality. It seems that although some of the generalised correlations are significantly different from zero, none of them vary across the levels of the independent variable blocks.

```
gen_cor(x = response, y = pesticide, U = 3, V = 3, rounding = 3)

##
##          Generalised Correlations
##
## data:  response, pesticide
##
## Table of correlations and p-values
## w = 0
##                 V_11  V_12  V_13  V_21   V_22  V_23  V_31
## correlations  0.354 0.683 0.347 0.400 -0.114 0.072 0.116
## chi-squared   0.061 0.000 0.067 0.034  0.547 0.703 0.539
## t-test        0.047 0.000 0.008 0.021  0.559 0.682 0.562
## Wilcoxon      0.059 0.001 0.007 0.026  0.600 0.305 0.699
## Shapiro-Wilk  0.275 0.697 0.241 0.789  0.563 0.043 0.708
##                 V_32   V_33
## correlations  0.048 -0.215
## chi-squared   0.800  0.256
## t-test        0.812  0.193
## Wilcoxon      0.909  0.158
## Shapiro-Wilk  0.154  0.961

np_anova(ordered_vars = data.frame(response,pesticide),
         predictor_vars = data.frame(block), uvw = c(3,1))
```

```
## Anova Table (Type III tests)
##
## Response: ortho_poly_response
##              Sum Sq Df F value  Pr(>F)
## (Intercept) 0.0090768  1  7.7715 0.01022 *
## block       0.0095566  3  2.7274 0.06634.
## Residuals   0.0280310 24
## ---
## Signif. codes:  0 '***' 0.001 '**' 0.01 '*' 0.05 '.' 0.1 ' ' 1
```

10.2   Ordered NP ANOVA for the CMH Design

As in Chapter 8, it makes sense to introduce ordered NP ANOVA by constructing these tests for the CMH design, and then extending to other designs. The low-order NP ANOVA tests for other designs, such as the Latin squares design, will be useful tests in their own right, and will be considered in section 10.5.

To construct the new tests, it is necessary to use results about partitioning Pearson's X_P^2 statistic for three-way tables of counts. In the context of this chapter, responses and treatments are ordered, while strata are not. This affects the partitioning.

Again, to more conveniently import the results needed, instead of using the notation N_{ihj} as in the CMH material, we will use N_{rsj} with r for response instead of h, and with the response being the first instead of the second subscript. As treatments are considered to be ordered here, in the counts the subscript i is changed to s to reflect that change.

An outline of the procedure of the NP ANOVA for ordered treatments is as follows:

- We wish to analyse categorical data from a particular ANOVA.
- The data can be presented as a table of counts such as $\{N_{rsj}\}$ in the next section.
- By partitioning Pearson's X_P^2 statistic for this table, we obtain components Z_{uvj}.
- Suppose $a_1(r), a_2(r), \ldots$ are orthonormal polynomials on $\left\{\frac{N_{r\bullet\bullet}}{N_{\bullet\bullet\bullet}}\right\}$ and $b_1(s), b_2(s), \ldots$ are orthonormal polynomials on $\left\{\frac{N_{\bullet s\bullet}}{N_{\bullet\bullet\bullet}}\right\}$. For a given pair (u, v), the components $\{Z_{uvj}\}$ are essentially the same as $\{a_u(r) b_v(s)\}$.
- Under a smooth model, the $\{a_u(r) b_v(s)\}$ for different pairs (u, v) are uncorrelated, and these responses at least asymptotically have other useful properties.

- Provided the necessary assumptions are satisfied, the modified ANOVA of particular interest, or other appropriate analysis, may be applied to sets of $\{a_u(r) b_v(s)\}$, perhaps for u and $v = 1, 2, 3$. As the $\{a_u(r) b_v(s)\}$ of different pairs (u, v) are uncorrelated, inference for any particular pair is not affected by inference for any other pair. The ANOVA with $u = 1$ and $v = 1$ gives an assessment of correlation in the familiar sense. The ANOVAs with $u = 1$ and $v = 2$ and with $u = 2$ and $v = 1$ assess umbrella effects. These and further generalised correlations may be used to construct models of conditional moments.
- It is desirable to confirm the assumptions of each ANOVA. Typically, we don't assess the smooth model; all that is lost if the smooth model is invalid is that inferences of different orders are no longer uncorrelated. If the ANOVA assumptions are dubious resampling p-values can be calculated. If that is not convenient, the robustness of the ANOVA suggests the p-values are generally reliable.

Subsequently, we will refer to the analysis based on $\{a_u(r) b_v(s)\}$ as the ordered NP ANOVA of degree (u, v).

10.3 Doubly Ordered Three-Way Tables

The ordered NP ANOVA will now be developed for the CMH design. However, as with the unordered NP ANOVA, we will change the notation slightly. The data are intended to be analysed by a fixed-effects ANOVA with two factors, treatments and blocks. These data are categorical responses specified by a table of counts $\{N_{rsj}\}$ in which r indexes the responses and s indexes the treatments; both are assumed to be ordinal. As usual j indexes blocks, or strata.

For the table of counts $\{N_{rsj}\}$ Pearson's statistic X_P^2 is given by $X_P^2 = \sum_{u=1}^{n-1} \sum_{v=1}^{t-1} \sum_{j=1}^{b} Z_{uvj}^2$ in which the Z_{uvj}, the components, are given by

$$Z_{uvj} = \sqrt{\frac{n}{p_{\bullet\bullet j}}} \sum_{r=1}^{n} \sum_{s=1}^{t} a_u(r) b_v(s) p_{rsj} = \frac{1}{\sqrt{n_{\bullet\bullet j}}} \sum_{r=1}^{n} \sum_{s=1}^{t} a_u(r) b_v(s) n_{rsj}$$

in which $p_{rsj} = \frac{n_{rsj}}{n}$ with $n = n_{\bullet\bullet\bullet}$. There are $(nt - 1)(b - 1)$ degrees of freedom: the number of unconstrained θs in the model given subsequently. For details see Rayner and Beh (2009).

The $(1, 1)$th component is $Z_{11j} = \sum_{r=1}^{n} \sum_{s=1}^{t} \frac{(r - \mu_1)(s - \mu_2) n_{rsj}}{\sigma_1 \sigma_2 \sqrt{n_{\bullet\bullet j}}}$, which is $\sqrt{n_{\bullet\bullet j}}$ times the Pearson correlation.

Comparing Z_{uvj} and V_{uvj}, it is apparent that Z_{uvj} is proportional to the (u, v)th generalised correlation test statistic for the table $\{N_{rsj}\}$ for a given j.

For a given (u, v) pair and a given observation there is only one set of outcomes (r, s, j) and for this observation $n_{rsj} = 1$, while for all other triples (r, s, j), $n_{rsj} = 0$. Thus, $\{Z_{uvj}\} = \{a_u(r)b_v(s)\}$. One way to assess the value of the components, and what information they give about the ANOVA of interest, is to apply that ANOVA to $\{Z_{uvj}\}$. This analysis is only valid if the ANOVA assumptions for each data set are satisfied. An equivalent parametric analysis is to apply the one-sample t-test to the data, provided of course, the data are consistent with normality. The signed-rank test may be applied if they are not.

An appropriate model is that for each r, s pair the counts follow a multinomial distribution with total count $n_{\bullet\bullet j}$ and probabilities

$$p_{rsj} = p_{rs\bullet} \sum_{u=0}^{n-1} \sum_{v=0}^{t-1} \theta_{uvj} a_u(r) b_v(s) \quad \text{for } j = 1, \ldots, b,$$

in which $\theta_{00j} = 1$, and $\theta_{u0j} = \theta_{0vj} = 0$. Assuming this model, under the null hypothesis that all θ_{uvj} (not defined to be 0 or 1) are zero, the Zs asymptotically have mean zero, variance one and for different (u, v) pairs are uncorrelated. If the θ_{uvj} are assumed to be of order $n^{-0.5}$ the same is asymptotically true in general. The details are similar to those in Rayner and Best (2013, Appendix B).

Since the ANOVA of interest is a fixed-effects model, and as the orthonormal functions are assumed to be polynomials, the ANOVA tests hypotheses about the expected response, $E\left[a_u(R)b_v(S)\right]$, and these involve bivariate moments up to degree (u, v).

In Section 8.3, we show that for the CRD and RBD the first-degree unordered NP ANOVA tests coincided with the ANOVA F tests that were equivalent to the Kruskal–Wallis and Friedman tests. The parallel questions here are how are the degree $(1, 1)$ ordered NP ANOVA tests related to the CMH C test? In Section 2.5.2, we gave the result that when there is only one stratum, $S_C = (n-1)r_P^2$. That is essentially the case here, and r_P, the Pearson correlation, is essentially the degree $(1, 1)$ generalised correlation. That being the case, we can test if the degree $(1, 1)$ generalised correlation is consistent with zero using the one-sample t-test or the signed rank test, or using S_C and the χ_1^2 distribution. The following examples show a satisfactory agreement among the competing approaches. However, beyond the CRD scenario, we are unable to find a relationship between S_C and the degree $(1, 1)$ ordered NP ANOVA test.

10.3.1 Whiskey Data Revisited

For the Whiskey example, in Section 2.5.3 a χ_1^2 p-value of 0.0494 was reported for the CMH C test. In testing for the degree (1, 1) generalised correlation being zero, the one sample t-test gave a p-value of 0.0418, while the signed rank test gave a p-value of 0.0355. The Shapiro–Wilk test for normality had a p-value of 0.0600 indicating that normality may be in doubt; therefore, one may have marginal preference for the latter p-value.

```
gen_cor(x = grade, y = maturity, U = 1, V = 1,
         x_scores = c(1,2,3), y_scores = c(1,5,7))

## 
##          Generalised Correlations
## 
## data:   grade, maturity
## 
## Table of correlations and p-values
## w = 0
## correlations  chi-squared      t-test     Wilcoxon Shapiro-Wilk
##      -0.7428       0.0356      0.0418       0.0355        0.0600
```

10.3.2 Jams Data Revisited

In Section 2.2, a χ_1^2 p-value 0.2936 was reported for the CMH C test. In testing for the degree (1, 1) generalised correlation being zero, the one sample t-test gave a p-value of 0.2249, while the signed rank test gave a p-value of 0.5765. However, the Shapiro–Wilk test for normality had a p-value of 0.0046, so the latter has more credibility.

```
gen_cor(x = sweetness, y = type, U = 1, V = 1)

## 
##          Generalised Correlations
## 
## data:   sweetness, type
## 
## Table of correlations and p-values
## w = 0
## correlations  chi-squared      t-test     Wilcoxon Shapiro-Wilk
##       0.2177       0.2861      0.2249       0.5765        0.0046
```

10.4 Extension to Other Designs

Consider again ANOVA designs consistent with the general linear model. The ordered NP ANOVA assumes that at least one of the independent variables is ordered in the sense that the levels of the factors corresponding to that variable are ordered.

Without loss of generality, suppose we have m independent variables with the first t of these being ordered while the remaining $m - t$ variables are not. Here $t = 1, \ldots, m$ and $m = 2, 3, \ldots$.

The unordered independent variables are combined and indexed as in the unordered case in Chapter 8. Suppose that $N_{rs_1 \ldots s_t i}$ counts the number of times the rth ordered response is assigned to the $(s_1, \ldots, s_t, i)$th combination of the ordered variables and the ith set of independent variables. The total number of observations is $n = N_{\bullet \ldots \bullet}$.

Write $p_{rs_1 \ldots s_t i} = \frac{N_{rs_1 \ldots s_t i}}{n}$ and construct sets of orthonormal polynomials, $\{a_u(r)\}$ on $p_{r \bullet \ldots \bullet}$ and for $w = 1, \ldots, t$, $\{a_{v_w}(s_j)\}$ on $\{p_{\bullet \ldots \bullet s_j \bullet \ldots \bullet}\}$. The polynomials of degree zero are all taken to be identically one. For a given combination of the response $\{a_u(r)\}$ and the levels $\{v_1, \ldots, v_t\}$ of the factors with ordered levels, define

$$Z_{uv_1 \ldots v_t i} = \sum_r \sum_{s_1} \cdots \sum_{s_t} a_u(r) a_{v_1}(s_1) \ldots a_{v_t}(s_t) \frac{N_{rs_1 \ldots s_t i}}{\sqrt{n}}.$$

In Rayner (2017, Appendix A), it is shown that for the table $\left\{ N_{rs_1 \ldots s_t i} \right\}$ the sum of the squares of the $Z_{uv_1 \ldots v_t i}$ is X_P^2. The statistic X_P^2 may be used to test for independence in $\left\{ N_{rs_1 \ldots s_t i} \right\}$, but that is not of interest here. The partition is arithmetic and does not depend on a model. Essentially, the information in the table $\left\{ N_{rs_1 \ldots s_t i} \right\}$ is also in the $\left\{ Z_{uv_1 \ldots v_t i} \right\}$. Partitioning the information in this table allows us to focus on the most interesting information, that corresponding to each of $u, v_1, \ldots, v_t$ that individually are small, namely 0, 1, or 2.

Recall that for a given observation, $N_{rs_1 \ldots s_t i}$ is the number of times the rth response is assigned to this set of values of ordered and independent variables. For any particular observation, there is just one set $\{r, s_1, \ldots, s_t, i\}$, and so for this observation, the count $N_{rs_1 \ldots s_t i}$ is one with counts for other sets being zero. Thus, for this observation $Z_{uv_1 \ldots v_t i} = \frac{a_u(r) a_{v_1}(s_1) \ldots a_{v_t}(s_t)}{\sqrt{n}}$. As we run through all possible observations, we find $\left\{ Z_{uv_1 \ldots v_t i} \right\} = \left\{ \frac{a_u(r) a_{v_1}(s_1) \ldots a_{v_t}(s_t)}{\sqrt{n}} \right\}$.

We now give a smooth model for the table of counts $N_{rs_1 \ldots s_t i}$. For the $(v_1, \ldots, v_t)$th combination of the ordered independent variables define a multinomial with parameters 1 and $\left\{ p_{rs_1 \ldots s_t} \right\}$, where

$$p_{rs_1 \ldots s_t} = p_{r \bullet \ldots \bullet} p_{\bullet s_1 \bullet \ldots \bullet} \ldots p_{\bullet \ldots \bullet s_t \bullet} \sum_u \sum_{v_1} \cdots \sum_{v_t} \theta_{uv_1 \ldots v_t} a_u(r) a_{v_1}(s_1) \ldots a_{v_t}(s_t).$$

The $\theta_{uv_1 \ldots v_t}$ are generalised correlations as described previously. It can be shown that $E\left[Z_{uv_1 \ldots v_t i}\right]$ is proportional to $\theta_{uv_1 \ldots v_t}$. Moreover, the Zs can be shown to be efficient score statistics and appropriate test statistics for testing hypotheses about the $\theta_{uv_1 \ldots v_t}$. See Rayner and Beh (2009).

As in the unordered case, every $Z_{uv_1 \ldots v_t i}$ is uncorrelated with every other Z. Because the $Z_{uv_1 \ldots v_t i}$ for different sets $(u, v_1, \ldots, v_t)$ are uncorrelated, an analysis indexed by $(u, v_1, \ldots, v_t)$ may be performed using the $\left\{ Z_{uv_1 \ldots v_t i} \right\}$, or, equivalently, the $\left\{ a_u(r) a_{v_1}(s_1) \ldots a_{v_t}(s_t) \right\}$: the analysis may be applied to $\left\{ a_u(r) a_{v_1}(s_1) \ldots a_{v_t}(s_t) \right\}$ for different choices of $(u, v_1, \ldots, v_t)$. Appropriate choices would often result in assessing bivariate generalised correlations of low degree, such as θ_{11} and θ_{21}.

As previously, it is convenient to refer to the analysis for a particular choice of $(u, v_1, \ldots, v_t)$ as being of that degree. For all particular choices of degree $(u, v_1, \ldots, v_t)$, to test for all $\theta_{uv_1 \ldots v_t} = 0$, each against $\theta_{uv_1 \ldots v_t} \neq 0$, re-parametrise to the parameters of the class of ANOVAs appropriate for the design. For example, if the ANOVA was an m-way factorial model with replication, then use the $(m - t)$-way factorial model with main and interaction effects up to order $m - t$. Apply that ANOVA – or other appropriate analysis – to $\left\{ a_u(r) a_{v_1}(s_1) \ldots a_{v_t}(s_t) \right\}$.

For some problems, it may be helpful to ignore the ordering for some factors with ordered levels, to assess less-complex, low-degree generalised correlations. In general, we recommend the use of permutation test p-values, and the procedure can be fairly described as non-parametric, depending only on the mild assumption of the smooth multinomial model. However, if it is inconvenient to use permutation test p-values, then p-values based on ANOVA F tests are generally reliable.

We now consider the ordered NP ANOVA for the LSD.

10.5 Latin Square Rank Tests

10.5.1 Doubly Ordered Four-Way Tables

For the table of counts $\{N_{rsjk}\}$ with both r and s ordinal, define the components

$$
\begin{aligned}
Z_{uvjk} &= \sqrt{\frac{n}{p_{\bullet\bullet j \bullet} p_{\bullet\bullet\bullet k}}} \sum_{r=1}^{n} \sum_{s=1}^{t} a_u(r) b_v(s) p_{rsjk} \\
&= \sqrt{\frac{n}{n_{\bullet\bullet j \bullet} n_{\bullet\bullet\bullet k}}} \sum_{r=1}^{n} \sum_{s=1}^{t} a_u(r) b_v(s) n_{rsjk}.
\end{aligned}
$$

As in Rayner and Beh (2009), Pearson's statistic X_P^2 is given by $X_P^2 = \sum_{u=1}^{n-1} \sum_{v=1}^{t-1} \sum_{j=1}^{b} \sum_{k=1}^{t} Z_{uvjk}^2$. A degree (u, v) assessment could be based on $\sum_{j=1}^{b} \sum_{k=1}^{t} Z_{uvjk}^2$ or perhaps $\left(\sum_{j=1}^{b} \sum_{k=1}^{t} Z_{uvjk} \right)^2$.

For each (u, v) and any observation, there is only one (r, s) pair for which n_{rsjk} is non-zero, and then it is one. This corresponds to precisely one row/column pair, although there may be other pairs with the same response/treatment pair. Thus for each observation, there is a pair (r, s) and a pair (j, k) for which $z_{uvjk} = a_u(r) b_v(s) \sqrt{\frac{n}{n_{\bullet\bullet j \bullet} n_{\bullet\bullet\bullet k}}}$, and hence,

$$\{Z_{uvjk}\} = \left\{ a_u(r) b_v(s) \sqrt{\frac{n}{n_{\bullet\bullet j \bullet} n_{\bullet\bullet\bullet k}}} \right\}.$$

Latin Square Design For Latin square designs, $n_{\bullet s \bullet\bullet} = n_{\bullet\bullet j \bullet} = n_{\bullet\bullet\bullet k} = t$ and $n = t^2$, so that $Z_{uvjk} = \sum_{r=1}^{n} \sum_{s=1}^{t} a_u(r) b_v(s) n_{rsjk}$. If the responses are ranks, and if ranking is overall, from 1 to t^2 and with no ties, then put $\mu_1 = \frac{t^2+1}{2}, \sigma_1^2 = \frac{(t^2-1)(t^2+1)}{12}, \mu_2 = \frac{t+1}{2}, \sigma_2^2 = \frac{(t-1)(t+1)}{12}$. Then, for example,

$$Z_{11jk} = \sum_{r=1}^{n} \sum_{s=1}^{t} \frac{(r - \mu_1)(s - \mu_2) n_{rsjk}}{\sigma_1 \sigma_2}.$$

As with the unordered ANOVA, we select a pair (u, v) corresponding to (one of) the generalised correlation(s) of interest for the data at hand. As we run through the observations, we run through all pairs (j, k), so that $\{Z_{uvjk}\} = \{a_u(r) b_v(s)\}$. These become the responses or treatments in a modified two-way ANOVA. The null hypothesis is that the treatments, that is the (u, v)th generalised correlations, do not vary over the levels of the independent variables.

We may also test the null hypothesis that the same generalised correlation, aggregated over the independent variables, is zero. Effectively, we have independent and identically distributed random variables W_j, say, $j = 1, \ldots, n$, and we wish to test $E[W] = 0$ against not this. Here $E[W]$ is the (u, v)th generalised correlation, and the hypothesis testing may be done using the t-test if the data are consistent with normality, or the signed rank test otherwise.

Traffic Data Revisited The analysis of the raw data found that the treatments, the signal sequences, were significant at the 0.05 level. The unordered NP ANOVA analysis of the ranks found significant first-order effects, but that the second- and third-order effects were not significant.

We now assume that the treatments are ordered from A to E. Nine generalised correlations with degrees from (1, 1) to (3, 3) were tested for consistency with zero, and only one, the (3, 2)th, had p-value less than 0.1.

In this case, the *p*-values were 0.0687 for the *t*-test and 0.0900 for the signed rank test. Since the Shapiro–Wilk test returned a *p*-value of 0.5411, there is no reason to doubt the *t*-test. However, since we have performed nine tests of significance, it would not be surprising if one of them be found significant even if there is no effect.

Next, we performed two-way ANOVAs for the nine generalised correlations with indexes from (1, 1) to (3, 3). The only effect of interest was for rows for the (2, 1)th generalised correlation, where a *p*-value of 0.0562 was found. This may also be an artefact of multiple testing. There is no compelling evidence that any of the generalised correlations vary across rows and columns.

```
gen_cor(x = rank(minutes), y = treatment, U = 3, V = 3)

##
##          Generalised Correlations
##
## data:   rank(minutes), treatment
##
## Table of correlations and p-values
## w = 0
##                   V_11    V_12    V_13    V_21    V_22    V_23    V_31
## correlations   -0.0628  0.1790  0.0863  0.0256  0.0453  0.0463 -0.1003
## chi-squared     0.7537  0.3708  0.6661  0.8981  0.8207  0.8168  0.6159
## t-test          0.7645  0.3816  0.6764  0.8986  0.8290  0.8150  0.6239
## Wilcoxon        0.7474  0.5294  0.6891  0.8519  0.8929  0.9357  0.3603
## Shapiro-Wilk    0.6094  0.2530  0.2935  0.1241  0.4773  0.1068  0.0694
##                   V_32    V_33
## correlations   -0.3357  0.2437
## chi-squared     0.0932  0.2231
## t-test          0.0687  0.2843
## Wilcoxon        0.0900  0.3673
## Shapiro-Wilk    0.5411  0.1601


np_anova(ordered_vars = data.frame(rank(minutes),treatment),
         predictor_vars = data.frame(intersection,time_of_day),
         uvw = c(2,1))

## Anova Table (Type III tests)
##
## Response: ortho_poly_response
##                  Sum Sq Df F value  Pr(>F)
## (Intercept)   0.0000407  1  0.0321 0.86001
## intersection  0.0146510  4  2.8902 0.05619.
## time_of_day   0.0030201  4  0.5958 0.67084
## Residuals     0.0202767 16
## ---
## Signif. codes:  0 '***' 0.001 '**' 0.01 '*' 0.05 '.' 0.1 ' ' 1
```

10.6 Modelling the Moments of the Response Variable

In the ordered NP ANOVA, most users will only be comfortable interpreting bivariate (u, v)th generalised correlations with u and $v = 0$, 1, and 2. However generalised correlations may also be used to model the moments of the response variable. To demonstrate the method and its drawbacks, here attention is restricted to consideration of an ordered ANOVA in which only the response and a single independent variable are ordered. All other variables have been summed out.

Consider the following smooth model for the counts:

$$p_{rs} = p_{r\bullet} p_{s\bullet} \sum_{u=0}^{n-1} \sum_{v=0}^{t-1} \theta_{uv} a_u(r) b_v(s)$$

in which $\theta_{00} = 1$, and $\theta_{u0} = \theta_{0v} = 0$. The probability function for the response conditional on the ordered independent variable taking a particular value s, $Y|s$, is $p_{r\bullet} \sum_{u=0}^{n-1} \sum_{v=0}^{t-1} \theta_{uv} a_u(r) b_v(s)$. It follows that for $u' \geq 1$

$$E\left[a_{u'}(Y)\,|s\right] = \sum_{r=1}^{n} a_{u'}(r) p_{r\bullet} \sum_{u=0}^{n-1} \sum_{v=0}^{t-1} \theta_{uv} a_u(r) b_v(s)$$

$$= \sum_{u=0}^{n-1} \sum_{v=0}^{t-1} \theta_{uv} b_v(s) \sum_{r=1}^{n} a_u(r) a_{u'}(r) p_{r\bullet}$$

$$= \sum_{v=0}^{t-1} \theta_{u'v} b_v(s).$$

Now, since $a_1(y) = \frac{y - \mu}{\sigma}$

$$E[Y|s] = \mu + \sigma E\left[a_1(Y)\,|s\right] = \mu + \sigma \left\{ \sum_{v=0}^{t-1} \theta_{1v} b_v(s) \right\}.$$

If there is no first-order effect, all θ_{1v} will be zero, and the conditional means are all equal to the response mean; otherwise, $E[Y|s] \neq E[Y]$.

Next, consider

$$a_2(y) = \frac{\left[(y - \mu)^2 - \frac{\mu_3(y - \mu)}{\mu_2} - \mu_2\right]}{\sqrt{d}}$$

in which $d = \mu_4 - \frac{\mu_3^2}{\mu_2} - \mu_2^2$, where the moments are of the responses. Now

$$E\left[a_2(Y)\,|s\right] \sqrt{d} = E\left[(Y - \mu)^2 |s\right] - \frac{\mu_3 E\left[(Y - \mu)\,|s\right]}{\mu_2} - \mu_2$$

from which

$$\text{Var}(Y|s) = \text{E}\left[(Y - \mu)^2|s\right] - (\text{E}[Y|s] - \mu)^2$$

$$= \mu_2 + \mu_3 \frac{\text{E}[(Y - \mu)|s]}{\mu_2} + \sqrt{d}\,\text{E}\left[a_2(Y)|s\right] - (\text{E}[Y|s] - \mu)^2$$

$$= \mu_2 + \mu_3 \sum_{v=0}^{t-1} \frac{\theta_{1v} b_v(s)}{\sigma} + \sqrt{d} \sum_{v=0}^{t-1} \theta_{2v} b_v(s)$$

$$- \mu_2 \left[\sum_{v=0}^{t-1} \theta_{1v} b_v(s)\right]^2 .$$

If there are no first- and second-order effects, that is all θ_{1v} and all θ_{2v} are zero, the conditional variances are all μ_2. The test for second-order effects is assessing whether or not the $a_2(y)$, and hence, a particular linear combination of moments up to order two, are consistent across levels.

In a similar vein if there are no effects to third order, then the test for third-order effects is assessing whether or not third moments are consistent across levels. Otherwise, the test for third-order effects is assessing a third-order effect in the moments, whether or not a particular linear combination of moments up to order three is consistent across levels.

These considerations of conditional moments do not apply to the unordered analysis. It seems then that the table $\{N_{ri}\}$ is so sparse that useful estimation is only possible if, as in the ordered analysis, further structure can be assumed.

To reflect on the usefulness or otherwise of these moment estimates, we consider again the strawberry data.

Strawberry Data Revisited The pesticide means are 142.4, 144.4, 103.6, 123.8, and 174.625. The corresponding modelled means using $\text{E}[Y|s] = \mu + \sigma\left[\theta_{1v} b_v(s)\right]$ and the three (near) significant generalised correlations of order (1, 1), (1, 2), (1, 3) are 144.3, 136.7, 115.2, 116.1, and 175.8. See Figure 10.2(a).

```
# true means
tapply(X = response, INDEX = pesticide, FUN = mean)

##        A       B       C       D       O
## 142.400 144.400 103.600 123.800 174.625

# modelled means
mu = mu_r(x = response, r = 1)
sigma = sqrt(mu_r(x = response, r = 2))
theta_uv = gen_cor(x = response, y = pesticide,
                   U = 3, V = 3)$correlations[,,1]
b_v = t(orthogonal_scores(pesticide,1))
```

```
for (i in 2:3) b_v = rbind(b_v,t(orthogonal_scores(pesticide,i)))

mu + sigma*colSums(theta_uv[1,1:3]*b_v[1:3,])

## [1] 144.3271 136.6916 115.1627 116.0916 175.8294
```

The pesticide variances are 364.24, 251.44, 23.04, 134.96, and 176.984. The corresponding modelled variances using Var $(Y|s)$ and the generalised correlations of orders $(1, 1), (1, 2), (1, 3)$, and $(2, 1)$ result in modelled variances of 391.2, 546.7, 8.5, 250.8, and 73.9. The pesticide B variance is poorly modelled, suggesting that there are highly significant generalised correlations involving the orthonormal polynomials of order four. See Figure 10.2(b).

```
# true variances
tapply(X = response, INDEX = pesticide, FUN = var)*
        c(rep(4/5,4),7/8)

##        A         B         C         D         O
## 364.2400 251.4400  23.0400 134.9600 176.9844

# modelled variances
mu_2 = mu_r(response,2)
sigma = sqrt(mu_2)
mu_3 = mu_r(response,3)
mu_4 = mu_r(response,4)
d = mu_4 - mu_3^2/mu_2 - mu_2^2
theta_uv = gen_cor(x = response, y = pesticide,
                   U = 3, V = 3)$correlations[,,1]
b_v = t(orthogonal_scores(pesticide,1))
for (i in 2:3) b_v = rbind(b_v,t(orthogonal_scores(pesticide,i)))

mu_2 + mu_3*colSums(theta_uv[1,1:3]*b_v[1:3,])/sigma +
    sqrt(d)*(theta_uv[2,1]*b_v[1,]) -
    mu_2*colSums(theta_uv[1,1:3]*b_v[1:3,])^2

## [1] 391.212315 546.686058   8.503665 250.751826  73.919409

gen_cor(x = response, y = pesticide,
        U = 3, V = 4)$output_table[,8,1]

## correlations  chi-squared    t-test      Wilcoxon Shapiro-Wilk
##   0.569437464 0.002585257 0.002907620 0.003307541 0.016699441
```

The poor agreement between the conditional variances and their estimates from the model suggests we take stock of the process here. The θ_{uv} in the modelled probability function are being estimated by their expectations, the sample generalised correlations. We could have used these estimates in the probability function to obtain density estimates. Instead, we took another step and estimated moments using the density estimates. Generally, there is no need to estimate these, as they are readily calculated directly. More sophisticated density and moment estimates are available, and no doubt some will use the sample generalised correlations, but perhaps not in as transparent a way as here.

One reason why the variance estimation may fail is that the orthonormal functions used, the polynomials, do not give a good low-order approximation to the probability function. This is discussed in a goodness

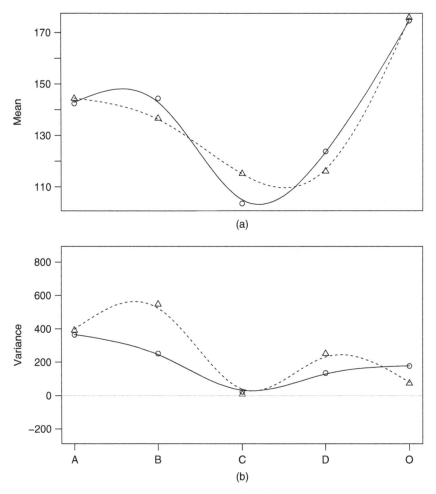

Figure 10.2 Exact (circles) pesticide means shown using with modelled (triangles) means (a); exact pesticide variances with modelled variances (b).

of fit setting in Rayner and Best (1986). The idea is that better estimation results from using orthonormal functions appropriate for the density.

In this setting, it would appear that if the moment estimates are poor, the orthonormal polynomials do not give a good low-order density estimate. In other words, if there are higher-degree components that are significant, these are required to give reasonable density and moment estimates. If the higher-order moments are not routinely calculated, then one way of being alerted to the need to calculate higher generalised correlations is to check the moment estimates. If they are obviously poor, there is information in the higher-order components.

10.7 Lemonade Sweetness Data

Five lemonades with increasing sugar content are ranked by each of 10 judges. They were not permitted to give tied outcomes. For this randomised block design, the data are in Table 10.3. We wish to assess what, if any, differences there are between the lemonades.

Routine calculation finds the Friedman statistic $F dm_A = 9.04$. The χ_4^2 p-value is 0.0601. There is no evidence, at the 0.05 level of a difference in lemonades. The F test statistic using the formula in Section 4.6 takes the value 2.6279 with $F_{4,36}$ p-value 0.0504. Further investigation, perhaps with more judges, would seem to be warranted.

The NP ANOVA ignoring order applies the randomised blocks ANOVA to the data transformed by the orthonormal polynomials of orders one, two, and three. The p-values are summarised in the Table 10.4. This ignored the ordering by sweetness of the lemonades. Normality is dubious so we calculated permutation test p-values based on 1,000,000 permutations

Table 10.3 Lemonade sweetness ranked by 10 judges.

Lemonade	Judge										Product mean
	1	2	3	4	5	6	7	8	9	10	
1	5	3	4	5	3	4	5	3	1	3	3.6
2	2	5	3	2	5	3	2	5	4	1	3.2
3	1	2	2	1	2	2	1	2	2	2	1.7
4	3	1	5	3	1	5	3	1	5	4	3.1
5	4	4	1	4	4	1	4	4	3	5	3.4

Table 10.4 Summary of the non-parametric unordered analysis for the lemonade data.

p-value	First order	Second order	Third order
F test – *F* distribution	0.0504	0.9232	0.0047
F test – permutation test	0.0533	0.9179	0.0063
Shapiro–Wilk Normality test	0.0211	<0.0001	0.0125

Table 10.5 Lemonade polynomial means.

Lemonade variety	1	2	3	4	5
First-order mean	0.424	0.141	−0.919	0.071	0.283
Second-order mean	−0.120	0.000	−0.060	0.299	−0.120
Third-order mean	−0.141	0.424	0.778	−0.141	−0.919

and found that the F distribution was an excellent approximation to the permutation distribution. It seems there is a weak first-order effect and a strong third-order effect – roughly indicating a mean effect and an effect due to moments up to third order.

The polynomial means for the lemonade varieties are given in Table 10.5. The first-order effects are almost significant at the 0.05 level. It appears this may be due to an umbrella effect: the means decrease then increase. See Figure 10.3. This can be confirmed by calculating the orthogonal contrasts of F. The linear, quadratic, and remainder p-values are 0.8108, 0.0882, and 0.0021. There is a non-linear effect that is more complex than quadratic.

The second-order mean differences are just natural variation: they are not significant at all reasonable levels. The third-order effect is significant at the 0.01 level. 'Eyeballing' the third order means in Table 10.5 suggests there is an umbrella effect. However, it is not clear that such an effect is useful in interpreting the data.

Next, the ordering of the lemonades is not ignored. The responses are transformed using the orthonormal polynomial of degree r and the lemonades are transformed using the orthonormal polynomial of degree s. These are multiplied giving a degree (r, s) response for each judge. This is treated as a random sample and tested for an expectation of zero using the t-test if the data are consistent with normality and the signed rank test if they are not.

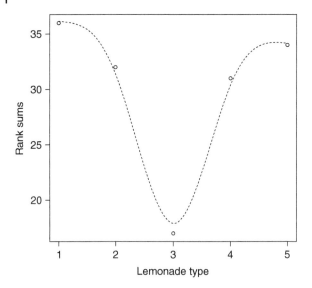

Figure 10.3 Lemonade rank sums versus lemonades.

None of the generalised correlations is consistent with normality, and therefore, testing for consistency with zero should be done using the signed-rank test. The generalised correlations of degree (1, 2), (3, 1), and (3, 2) are significant at the 0.05 level with *p*-values of 0.0067, 0.0194, and 0.0043, respectively. The remaining generalised correlations were consistent with being zero.

```
gen_cor(x = ranks, y = sugar_content, U = 3, V = 3, rounding = 3)

##
##          Generalised Correlations
##
## data:  ranks, sugar_content
##
## Table of correlations and p-values
## w = 0
##                 V_11   V_12   V_13   V_21    V_22    V_23    V_31
## correlations  -0.050  0.363  0.000  0.042  -0.079  -0.085  -0.300
## chi-squared    0.724  0.010  1.000  0.765   0.578   0.550   0.034
## t-test         0.720  0.007  1.000  0.772   0.570   0.579   0.026
## Wilcoxon       0.885  0.007  0.472  0.724   0.376   0.549   0.019
## Shapiro-Wilk   0.001  0.001  0.001  0.004   0.000   0.004   0.001
##                 V_32   V_33
## correlations  -0.473  0.050
## chi-squared    0.001  0.724
## t-test         0.001  0.696
## Wilcoxon       0.004  0.721
## Shapiro-Wilk   0.000  0.001
```

To assess whether any generalised correlation varied across levels of the blocks each was tested using a one-way ANOVA: the response was the generalised correlation and the treatment blocks. Those of significance or near significance were the (1, 3)th, with p-value 0.0250, and the (3, 1)th, with p-value 0.0757. We conclude that the (1, 2)th and (3, 2)th generalised correlations were different from zero but did not vary across blocks. The (3, 1)th generalised correlation was significantly different from zero and did vary across blocks. The (1, 3)th generalised correlation was not significantly different from zero but did vary across blocks. There is a complex correlation structure here.

```
np_anova(ordered_vars = data.frame(ranks,sugar_content),
         predictor_vars = data.frame(judge), uvw = c(3,1))

## Anova Table (Type III tests)
##
## Response: ortho_poly_response
##              Sum Sq Df F value  Pr(>F)
## (Intercept) 0.00128  1  4.3686 0.04301 *
## judge       0.00508  9  1.9264 0.07574 .
## Residuals   0.01172 40
## ---
## Signif. codes:  0 '***' 0.001 '**' 0.01 '*' 0.05 '.' 0.1 ' ' 1
```

The key conclusion here is that an initial increase in sugar content increases the acceptability of the lemonade, but increasing the sugar content beyond a certain point diminishes the lemonade's acceptability. The third product is the most acceptable. The significant generalised correlations indicate that a parabola doesn't best model the umbrella effect; a better model would use third-order terms. Some of the generalised correlations vary with the judges.

For these data, standard tools will identify which lemonades are different and identify the third product, with mean 1.7, as different from the other lemonades.

From the data, the means for lemonades 1 to 5 are 3.6, 3.2, 1.7, 3.1, and 3.4, respectively. Using E [$Y|s$] above and only the significant (1, 2) generalised correlation gives means of 3.61, 2.69, 2.39, 2.69, and 3.61. Including θ_{11} and θ_{13} as well as θ_{12} doesn't improve the fit. Of course, modelling E [$Y|s$] using θ_{11} to θ_{14} will give an exact fit, but that is not the point.

```
# true means
tapply(X = ranks, INDEX = sugar_content, FUN = mean)

##   1   2   3   4   5
## 3.6 3.2 1.7 3.1 3.4
```

```
# modelled means
mu = mu_r(x = ranks, r = 1)
sigma = sqrt(mu_r(x = ranks, r = 2))
theta_uv = gen_cor(x = ranks, y = sugar_content,
                  U = 4, V = 4)$correlations[,,1]
b_v = t(orthogonal_scores(sugar_content,1))
for (i in 2:4) {b_v = rbind(b_v,
      t(orthogonal_scores (sugar_content,i)))}

mu + sigma*theta_uv[1,2]*b_v[2,]

## [1] 3.614286 2.692857 2.385714 2.692857 3.614286
```

However, Figure 10.4 shows a much sharper curve than a parabola is needed to model the product means.

The variances for lemonades one to five are 1.44, 1.96, 0.21, 2.49, and 1.64, respectively. Using

$$\mathrm{Var}\,(Y|s) = \mu_2 + \mu_3 \sum_{v=0}^{t-1} \frac{\theta_{1v}b_v\,(s)}{\sigma} + \sqrt{d}\sum_{v=0}^{t-1} \theta_{2v}b_v\,(s) - \mu_2 \left[\sum_{v=0}^{t-1} \theta_{1v}b_v\,(s)\right]^2$$

including θ_{12} the variances are modelled as 1.62, 1.91, 1.62, 1.91, and 1.62. Including both θ_{12} and θ_{23} gives improved estimates 1.72, 1.71, 1.62, 2.11, and 1.52.

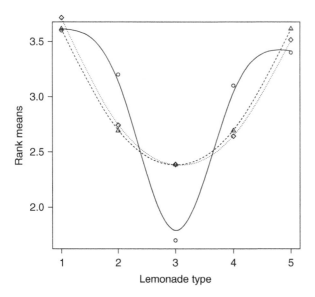

Figure 10.4 Exact means (circles) and models with just the (1, 2)th generalised correlation (triangles) and with the (1, 1), (1, 2), and (1, 3) generalised correlations (diamonds) versus lemonades.

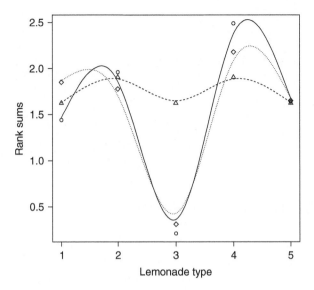

Figure 10.5 Exact variances (circles) and modelled variance using the (1, 2) generalised correlation (triangles) and with the (1, 2), (1, 4), and (2, 3) generalised correlations (diamonds) versus lemonades.

In Figures 10.4 and 10.5 none of the modelled moments using generalised correlations up to order three appear to track the conditional moments very well. We conclude that there is significant information in the remaining generalised correlations, and this is supported by the modelled variances that include the (1, 4) generalised correlation shown in Figure 10.5. This correlation is almost significant with a *p*-value of 0.09. The data analyst needs to judge whether or not this is relevant to understanding the data.

```
# true variances
tapply(X = ranks, INDEX = sugar_content, FUN = var)*rep(9/10,5)

##    1    2    3    4    5
## 1.44 1.96 0.21 2.49 1.64

# modelled variances
mu_2 = mu_r(ranks,2)
sigma = sqrt(mu_2)
mu_3 = mu_r(ranks,3)
mu_4 = mu_r(ranks,4)
d = mu_4 - mu_3^2/mu_2 - mu_2^2
theta_uv = gen_cor(x = ranks, y = sugar_content,
                   U = 4, V = 4)$correlations[,,1]
b_v = t(orthogonal_scores(sugar_content,1))
for (i in 2:4) {b_v = rbind(b_v,
     t(orthogonal_scores (sugar_content,i)))}
```

```
mu_2 + mu_3*(theta_uv[1,2]*b_v[2,])/sigma -
  mu_2*(theta_uv[1,2]*b_v[2,])^2
```

```
## [1] 1.622653 1.905663 1.622653 1.905663 1.622653
```

```
mu_2 + mu_3*(theta_uv[1,2]*b_v[2,])/sigma +
  sqrt(d)*(theta_uv[2,3]*b_v[3,]) -
  mu_2*(theta_uv[1,2]*b_v[2,])^2
```

```
## [1] 1.722653 1.705663 1.622653 2.105663 1.522653
```

10.8 Breakfast Cereal Data Revisited

This example was considered in Section 5.3. The data are rankings of breakfast cereals in a BIBD. We found the ANOVA F test on the ranks has F statistic 11.5556 with p-value 0.0001, while $D_A = 14.8571$ with χ_4^2 p-value 0.0050.

The first, second, and third degrees NP ANOVA unordered p-values are 0.0001, 0.8962, and 0.9322, respectively. The corresponding Shapiro–Wilk p-values for the residuals for first-, second-, and third-degree analyses are 0.3255, 0.0321, and 0.0075, respectively. The first-degree p-value is, as expected, the same as the ANOVA F p-value. There is a strong first-degree effect but no evidence of a higher degree effect, even if the third-degree Shapiro–Wilk p-value makes that inference dubious.

We now assume that the cereals are ordered A < B < C < D < E, perhaps by sugar content. The only significant generalised correlation is that of order (1, 2) with both t-test and signed rank test p-values less than 0.0001. The Shapiro–Wilk test on the residuals has p-value 0.0027, so the t-test result would normally be set aside.

```
np_anova(ordered_vars = data.frame(rank),
         predictor_vars = data.frame(type, judge), uvw = 1)

## Anova Table (Type III tests)
##
## Response: ortho_poly_response
##                Sum Sq Df F value    Pr(>F)
## (Intercept) 0.13932  1  8.6688 0.0095232 **
## type        0.74286  4 11.5556 0.0001327 ***
## judge       0.12381  9  0.8560 0.5796333
## Residuals   0.25714 16
## ---
## Signif. codes:  0 '***' 0.001 '**' 0.01 '*' 0.05 '.' 0.1 ' ' 1
```

```
gen_cor(x = rank, y = type, z = judge,
        U = 3, V = 3, W = 0, rounding = 3)
```

```
##
##            Generalised Correlations
##
## data:  rank, type, and judge
##
## Table of correlations and p-values
## w = 0
##                  V_11    V_12  V_13  V_21  V_22    V_23    V_31
## correlations 0.000  -0.782 0.077 0.000 0.060 -0.076 -0.089
## chi-squared  1.000   0.000 0.673 1.000 0.741  0.679  0.626
## t-test       1.000   0.000 0.662 1.000 0.744  0.692  0.643
## Wilcoxon     0.861   0.000 0.338 0.918 0.829  0.673  0.578
## Shapiro-Wilk 0.085   0.003 0.113 0.145 0.003  0.392  0.251
##                  V_32    V_33
## correlations 0.208  -0.018
## chi-squared  0.254   0.922
## t-test       0.250   0.927
## Wilcoxon     0.658   0.992
## Shapiro-Wilk 0.000   0.100
```

The mean rankings for cereals A to E are 1.25, 2.42, 2.75, 2.25, and 1.33, respectively. A plot of response against cereal in Figure 10.6 shows a clear

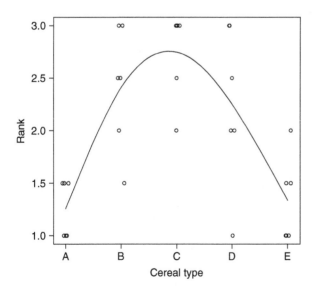

Figure 10.6   Ranks versus breakfast cereal type.

parabolic shape. If the ordering is based on sugar content, then tasters preferred mid-range sugar content.

For this design the ordered NP ANOVA involves using the observations of the various generalised correlations as responses and blocks as factor. The only significant result was for the (2, 3)th generalised correlation, indicating that these observations varied across judges. This apparent significance could be an artefact of there being approximately 5% significant results when performing multiple tests of significance at the 0.05 level. As all the other tests are not significant, it appears that the judges are acting homogeneously in their assessments of the cereals.

Bibliography

Rayner, J. C. W. (2017). Extended ANOVA. *Journal of Statistical Theory and Practice*, 11(1):208–219.

Rayner, J. C. W. and Beh, Eric J. (2009). Components of Pearson's statistic for at least partially ordered m-way contingency tables. *Advances in Decision Sciences*, 2009:9, Article ID 980706. https://doi.org/10.1155/2009/980706.

Rayner, J. C. W. and Best, D. J. (1986). Neyman-type smooth tests for location-scale families. *Biometrika*, 73(2):437–446.

Rayner, J. C. W. and Best, D. J. (2013). Extended ANOVA and rank transform procedures. *Australian & New Zealand Journal of Statistics*, 55(3):305–319.

11

Conclusion

11.1 CMH or NP ANOVA?

In Section 1.1 we described, albeit briefly, the CMH and NP ANOVA tests. We tried to make it clear that CMH methods are appropriate tools that may be employed by data analysts to analyse categorical response data for a certain, albeit restricted, class of designs. Several chapters later, the reader now has a fuller understanding of both the CMH and NP ANOVA tests. For some, the CMH methodology is embedded as their instrument for analysis of the designs of prime interest to them. We hope that as a result of having read the preceding chapters, these readers will have a fuller understanding of the methodology and be able to apply the extensions described here.

However, the origins of the CMH tests lie more than half a century ago, and perhaps there are more appropriate tools available today. The later chapters put the case for what we have called NP ANOVA. Here we give some key points in a comparison of the two methodologies.

- *CMH uses conditional inference while NP ANOVA uses conventional unconditional inference.* Conditional inference is a dated paradigm, and unfamiliar to most modern-day users. This is not a problem to those who are only interested in the tools and their use. While the Kruskal–Wallis and Friedman tests are CMH tests, they have alternative derivations and their usefulness is indisputable.
- *Software is not widely available for either methodology.* Of course, there are sources, and this text provides one such. However, programming NP ANOVA in, for example, ®, is routine.
- *The CMH design is very restricted.* We have highlighted examples such as the strawberry example, where the design is the supplemented balanced, and designs, such as the balanced incomplete block and Latin square designs, where CMH methods are not available. However, the Pearson

An Introduction to Cochran–Mantel–Haenszel Testing and Nonparametric ANOVA, First Edition. J.C.W. Rayner and G. C. Livingston Jr.
© 2023 John Wiley & Sons Ltd. Published 2023 by John Wiley & Sons Ltd.

tests permit analysis similar to that achieved by the nominal CMH tests, and NP ANOVA gives analysis similar to that achieved by the ordinal CMH tests.

- *Ease of explanation.* The statistician would have no problem explaining either methodology as a 'black box' – 'this works with your data, and here are the p-values' – or superficially, 'these are location and correlation tests for ordered categorical response data, and here are the p-values'. However, at a deeper level there can be little difficulty in using the entrenched Pearson tests, and NP ANOVA just uses ANOVA on transformed data. While orthonormal polynomials may not be familiar to unsophisticated users, they just give assessments accounting for second, third, … moments. It is certainly true that a discussion of generalised correlations may challenge some users, but that problem confronts both methodologies. Basically they generalise umbrella effects.

On that last point, a brief description of NP ANOVA may be something like the following. *Unordered NP ANOVA* applies the ANOVA of interest to the responses transformed by orthonormal polynomials of degree one, two, three, and so on, to give uncorrelated analyses of moment effects of degree one, up to degree two, up to degree three, and so on. Degree two means first and second moments, degree three means first, second and third moments, and so on. *Ordered nonparametric ANOVA* constructs a new response based on multiplying together the responses and ordinal treatments both transformed by orthonormal polynomials. It applies this new response to a modified 'cut-down' ANOVA. This allows the analysis of generalised correlation effects. The most familiar of these are linear–linear and umbrella effects. These analyses are uncorrelated.

For the reader who has persevered with the text, one useful outcome is you now have nonparametric umbrella tests for the Latin square, balanced incomplete block, and other important designs.

We now look at two examples to which we apply the full range of tests developed in the preceding chapters.

11.2 Homosexual Marriage Data Revisited for the Last Time!

This example was previously considered in sections 1.4, 2.2, 2.5.3, 3.7, 7.4, 8.5. Agresti (2003) reported that the CMH GA, MS, and C tests, all using natural scores for education, religion and the proposition responses, had p-values less than 0.001.

Table 11.1 Homosexual marriage opinions p-values for conditional and unconditional CMH tests.

Conditional			Unconditional		
Statistic	Value	p-value	Statistic	Value	p-value
S_{OPA}	26.71	0.0008	T_{OPA}	27.09	0.0007
S_{GA}	19.76	0.0006	T_{GA}	20.68	0.0004
S_{MS1}	17.94	0.0001	T_{M1}	23.71	0.0000
S_{MS2}	1.98	0.3713	T_{M2}	2.40	0.3017
S_{C11}	16.83	0.0000	$T_C = \dfrac{V^2_{11\bullet}}{2}$	17.48	0.0000
S_{C12}	1.91	0.1671	$\dfrac{V^2_{12\bullet}}{2}$	2.35	0.1254
S_{C21}	1.12	0.2902	$\dfrac{V^2_{21\bullet}}{2}$	1.29	0.2570
S_{C22}	0.07	0.7919	$\dfrac{V^2_{22\bullet}}{2}$	0.03	0.8726

The complete set of p-values for both the conditional and unconditional, and extended CMH analyses, are given in Table 11.1.

There is close agreement between the conditional and unconditional p-values.

There is strong evidence of overall and average association, and of a first order location effect, but no evidence of a second order effect. The latter is confirmed by the unordered NP ANOVA, which gave p-values of less than 0.0001 for the first order effect and 0.3719 for the second order effect.

There is strong evidence of a linear–linear correlation, due almost entirely to the college stratum. When the strata are considered individually the linear–linear correlation for college had p-value 0.0001, while that for school was 0.1185. Collectively the linear–linear correlation p-value was 0.0000. No other correlation effect was significant.

In testing if the generalised correlations of degrees (1, 1), (1, 2), (2, 1), and (2, 2) are zero the Wilcoxon signed rank test two-tailed p-values were 0.0003, 0.0637, 0.0711, and 0.2525. In all cases the t-test was invalid due to non-normality. There is strong evidence that the linear–linear correlation is significantly different from zero. In addition there is a suggestion of umbrella correlations: the (1, 2)th and the (2, 1)th, are not quite different from zero.

Ordered ANOVA may be used to test if generalised correlations vary across levels. The p-values corresponding to degrees (1, 1), (1, 2), (2, 1), and (2, 2) are 0.1466, 0.2884, 0.6654, and 0.9838, respectively. Thus there is insufficient evidence that any of the generalised correlations vary across levels.

In summary there isn't obvious agreement between the CMH and NP ANOVA analyses. Nevertheless both agree there is a (1, 1) correlation although NP ANOVA indicates the correlations do not vary significantly between school and college. CMH doesn't assess effects across levels, so there is no formal disagreement here. The separate analyses of school and college are, however, informative.

NP ANOVA suggests the (1, 2) and (2, 1) generalised correlations are only marginally non-significant, while CMH is clear about the non-significance. Most of the rows in Table 1.2 tend to decrease then increase, albeit only slightly, so 'eyeballing' the data supports the NP ANOVA conclusion. However, the effect is slight. Both analyses agree these effects are not significant; it is not surprising that they vary a little in the strength of their conclusions.

There is strong evidence that the different religions respond differently to the proposition, and also as religious belief becomes increasingly liberal so does agreement with the proposition.

```
gen_cor(x = opinion, y = religion, z = education,
        U = 2, V = 2, W = 0)

##
##        Generalised Correlations
##
## data:  opinion, religion, and education
##
## Table of correlations and p-values
## w = 0
##                  V_11      V_12      V_21      V_22
## correlations   -0.3597    0.1058   -0.1194   -0.0257
## chi-squared     0.0000    0.2224    0.1687    0.7672
## t-test          0.0000    0.2300    0.1779    0.7601
## Wilcoxon        0.0003    0.0637    0.0711    0.2525
## Shapiro--Wilk   0.0000    0.0000    0.0000    0.0000


np_anova(ordered_vars = data.frame(opinion,religion),
         predictor_vars = data.frame(education), uvw = c(1,1))

## Anova Table (Type III tests)
##
## Response: ortho_poly_response
##               Sum Sq  Df  F value  Pr(>F)
## (Intercept) 0.0001708   1   3.2164  0.07521.
## education   0.0001132   1   2.1328  0.14657
## Residuals   0.0069553 131
## ---
## Signif. codes:  0 '***' 0.001 '**' 0.01 '*' 0.05 '.' 0.1 ' ' 1
```

11.3 Job Satisfaction Data

The data in Table 11.2, from Agresti (2003), relate job satisfaction and income in males and females. Gender induces two strata; the treatments are income with categories scored 3, 10, 20, and 35 while the response is job satisfaction with categories scored 1, 3, 4, and 5. Agresti reports p-values for the CMH general association, mean scores, and correlation tests as 0.3345, 0.0288, and 0.0131, respectively.

Extended CMH analysis was used to check the values given by Agresti and to provide a comparison with the nonparametric ANOVA analysis. See Table 11.3. The conditional and unconditional p-values are similar without being uniformly close. The same conclusions are reached.

The CMH analysis shows no general nor overall partial association between income level and satisfaction.

The CMH mean scores p-value was confirmed, while the second and third degree analyses found conditional p-values of 0.8981 and 0.9027. There is no evidence of second or third degree effects. The unconditional analysis agreed.

The CMH correlation p-value was confirmed but we also found that the female stratum had Pearson correlation 0.1614 with p-value 0.2037. The male stratum had Pearson correlation 0.3710 with p-value 0.0205. It seems

Table 11.2 Satisfaction and income in males and females. The satisfaction codes correspond, respectively, to: Very Dissatisfied; A Little Satisfied; Moderately Satisfied; and, Very Satisfied.

Gender	Income ($1000)	Job satisfaction				
		VD	ALS	MS	VS	Total
Female	< 5	1	3	11	2	17
	5 to 15	2	3	17	3	25
	15 to 25	0	1	8	5	14
	>25	0	2	4	2	8
	Total female	3	9	40	12	64
Male	< 5	1	1	2	1	5
	5 to 15	0	3	5	1	9
	15 to 25	0	0	7	3	10
	> 25	0	1	9	6	16
	Total male	1	5	23	11	40

Table 11.3 Job satisfaction p-values.

Conditional			Unconditional		
Statistic	**Value**	**p-value**	**Statistic**	**Value**	**p-value**
S_{OPA}	20.59	0.3005	T_{OPA}	21.05	0.2777
S_{GA}	10.20	0.3345	T_{GA}	11.52	0.2415
S_{MS1}	9.03	0.0288	T_{M1}	10.80	0.0129
S_{MS2}	0.59	0.8981	T_{M2}	1.69	0.6381
S_{C11}	6.1563	0.0131	$T_C = \dfrac{V_{11\bullet}^2}{2}$	6.5803	0.0103
S_{C12}	0.0707	0.7903	$\dfrac{V_{12\bullet}^2}{2}$	0.0005	0.9826
S_{C21}	1.5200	0.2176	$\dfrac{V_{21\bullet}^2}{2}$	2.2637	0.1324
S_{C22}	0.0192	0.8897	$\dfrac{V_{22\bullet}^2}{2}$	0.2337	0.6288

the overall p-value of 0.0131 was largely due to the male stratum. These are conditional p-values; the unconditional ones are similar. It seems that males are more inclined than females to have greater job satisfaction with greater income. Only the $(1, 1)$ correlation was significant.

```
# set the scores
a_ij = matrix(rep(c(3,10,20,35),2),nrow=4)
b_hj = matrix(rep(c(1,3,4,5),2),nrow=4)

CMH(treatment = income, response = satisfaction, strata = gender,
    a_ij = a_ij, b_hj = b_hj, cor_breakdown = FALSE)

##
##        Cochran Mantel Haenszel Tests
##
##                                S df p-value
## Overall Partial Association 20.590 18 0.30050
## General Association         10.200  9 0.33450
## Mean Score                   9.034  3 0.02884
## Correlation                  6.156  1 0.01309

# change the variables to include scores
satisfaction_num = as.numeric(satisfaction)
income_num = as.numeric(income)
for (i in 1:4) satisfaction_num[satisfaction==i] = b_hj[i,1]
for (i in 1:4) income_num[income==i] = a_ij[i,1]
```

```
b_hj1 = orthogonal_scores(x = satisfaction_num, degree = 1,
                          n_strata = 2)
a_ij1 = orthogonal_scores(x = income_num, degree = 1, n_strata = 2)
b_hj2 = orthogonal_scores(x = satisfaction_num, degree = 2,
                          n_strata = 2)
a_ij2 = orthogonal_scores(x = income_num, degree = 2, n_strata = 2)

CMH(treatment = income, response = satisfaction, strata = gender,
    a_ij = a_ij1, b_hj = b_hj2, test_OPA = FALSE, test_GA = FALSE,
    cor_breakdown = FALSE)
```

```
##
##        Cochran Mantel Haenszel Tests
##
##                 S df p-value
## Mean Score  0.59270  3  0.8981
## Correlation 0.07068  1  0.7903
```

```
CMH(treatment = income, response = satisfaction, strata = gender,
    a_ij = a_ij2, b_hj = b_hj1, test_OPA = FALSE, test_GA = FALSE,
    test_MS = FALSE, cor_breakdown = FALSE)
```

```
##
##        Cochran Mantel Haenszel Tests
##
##               S df p-value
## Correlation 1.52  1  0.2176
```

```
CMH(treatment = income, response = satisfaction, strata = gender,
    a_ij = a_ij2, b_hj = b_hj2, test_OPA = FALSE, test_GA = FALSE,
    test_MS = FALSE, cor_breakdown = FALSE)
```

```
##
##        Cochran Mantel Haenszel Tests
##
##                 S df p-value
## Correlation 0.01923  1  0.8897
```

```
unconditional_CMH(treatment = income, response = satisfaction,
                  strata = gender, U = 2, V = 2, a_ij = a_ij,
                  b_hj = b_hj)
```

```
##
##        Unconditional Analogues to the CMH Tests
##
##                 T df p-value
## T_OPA    2.105e+01 18 0.27670
## T_GA     1.152e+01  9 0.24150
## T_M1     1.080e+01  3 0.01286
## T_M2     1.695e+00  3 0.63810
## T_C11.   6.580e+00  1 0.01031
## T_C12.   4.735e-04  1 0.98260
## T_C21.   2.264e+00  1 0.13240
## T_C22.   2.337e-01  1 0.62880
```

Applying the unordered NP ANOVA, first a two factor ANOVA with interaction resulted in a non-significant interaction term. Subsequently a model without interaction was applied resulting in significant first order income effects, with no second and third degree effects. The Shapiro–Wilk normality p-values are less than 0.01. The income p-values for the first to third order analyses are 0.0261, 0.9000, and 0.9074.

Although normality of the residuals is suspect for all analyses, it appears that the only effect of substance is a first order income effect. For the responses the gender means are 13.45 for females and 21.63 for males. The means for income levels are 6.50 for less than \$5,000, 14.07 for \$5,000 to \$15,000, 16.10 for \$15,000 to \$25,000, and 21.26 for above \$25,000. It appears that the two higher and two lower means are similar. There is some evidence of differences in job satisfaction: the higher income levels are happier than the lower levels.

Ordered nonparametric ANOVA was used to test the null hypothesis that the generalised correlations up to the (3, 3)th are zero. The Wilcoxon signed rank test was used because in each case the data were found to have Shapiro–Wilk normality test p-values of less than 0.0001. For degree (1, 1) up to degree (3, 3) the p-values were 0.0341, 0.5178, 0.8327, 0.1881, 0.9017, 0.1028, 0.3443, 0.8837, and 0.4018, respectively. The only generalised correlation to give evidence of not being zero is that of degree (1, 1). There is some evidence that job satisfaction and income increase together. The ordered NP ANOVA, a one factor ANOVA with factor gender, has a p-value of 0.1935, so there is no evidence that the (1, 1) generalised correlation varies between males and females. This is the only NP ANOVA test in disagreement with the CMH tests.

It seems the insignificant general association is because it is testing for too general an alternative: generalised correlations up to order (3, 3). The insignificant higher degree correlations mask the significant (1, 1) correlation. Alternatively the CMH GA test could be viewed as assessing moment effects up to order three, with the second and third order effects masking the first order effect.

There is a very satisfactory agreement between the conditional, unconditional, and nonparametric ANOVA analyses.

```
# unordered NP ANOVA with interaction
sat_poly_1 = poly(x = satisfaction_num,degree = 1)[,1]
Anova(lm(sat_poly_1~income + gender + income:gender), type = 3)

## Anova Table (Type III tests)
##
## Response: sat_poly_1
##                 Sum Sq Df F value Pr(>F)
```

```
## (Intercept)     0.00994  1  1.0739 0.3027
## income          0.04062  3  1.4632 0.2295
## gender          0.00705  1  0.7617 0.3850
## income:gender 0.01410    3  0.5080 0.6777
## Residuals       0.88830 96

np_anova(ordered_vars = satisfaction_num,
         predictor_vars = data.frame(income,gender), uvw = 1)

## Anova Table (Type III tests)
##
## Response: ortho_poly_response
##              Sum Sq Df F value  Pr(>F)
## (Intercept) 0.02480  1  2.7212 0.10219
## income      0.08798  3  3.2175 0.02607 *
## gender      0.00013  1  0.0146 0.90402
## Residuals   0.90240 99
## ---
## Signif. codes:  0 '***' 0.001 '**' 0.01 '*' 0.05 '.' 0.1 ' ' 1

# ordered NP ANOVA
gen_cor(x = satisfaction_num, y = income_num, U = 3, V = 3,
        rounding = 3)

##
##          Generalised Correlations
##
## data:  satisfaction_num, income_num
##
## Table of correlations and p-values
## w = 0
##                  V_11    V_12    V_13   V_21   V_22    V_23
## correlations    0.262  -0.117  -0.124  0.037  0.015  -0.071
## chi-squared     0.008   0.234   0.205  0.705  0.875   0.469
## t-test          0.007   0.229   0.212  0.705  0.876   0.476
## Wilcoxon        0.034   0.518   0.833  0.188  0.902   0.103
## Shapiro--Wilk   0.000   0.000   0.000  0.000  0.000   0.000
##                  V_31    V_32    V_33
## correlations    0.053   0.058   0.023
## chi-squared     0.587   0.557   0.814
## t-test          0.596   0.540   0.815
## Wilcoxon        0.344   0.884   0.402
## Shapiro--Wilk   0.000   0.000   0.000

np_anova(ordered_vars = data.frame(satisfaction_num,income_num),
         predictor_vars = data.frame(gender), uvw = c(1,1))

## Anova Table (Type III tests)
##
## Response: ortho_poly_response
##                 Sum Sq  Df F value Pr(>F)
## (Intercept) 0.0001591   1  1.8609 0.1755
## gender      0.0001464   1  1.7130 0.1935
## Residuals   0.0087194 102
```

11.4 The End

In Section 2.1 we noted that Mantel and Haenszel (1959) were writing about retrospective studies. However, most users of the CMH methodology are not; for most the interest is in prospective studies. For those users the paradigm is that of conditional testing, and that is little used in modern statistics.

The traditional CMH methodology is a suite of tests, for overall partial association, general association, mean score, and correlation. These tests seek alternatives to the null hypothesis of no association in parameter spaces of dimensions $b(c-1)(t-1)$, $(c-1)(t-1)$, $(t-1)$ and 1, respectively. These are the degrees of freedom associated with the tests. The OPA test is omnibus, seeking non-association in a very broad parameter space with modest power. As the tests become more focused their parameter spaces are reduced and the power increases for alternatives in the reduced parameter spaces. If, for example, $b = c = t = 5$, the dimensions of the parameter spaces are 80, 16, 4, and 1. In moving from one test to the next in the sequence, at least three quarters of the parameter space are ignored. There may be more power in detecting alternatives in the reduced parameter space, but there is no power for most of the initial parameter space.

In recent times many data analysts would take an exploratory data analytic approach and start with an omnibus test, decompose it into components, and apply both the omnibus test and all its components to the data. With this approach there may be choices as to how to do the decomposition, but often knowledge of how the data were generated can help here.

We have glossed over the nominal CMH tests somewhat, suggesting that whatever they can do the Pearson tests can do. Moreover decomposition of the Pearson tests has a rich history going back to Lancaster (1949) and beyond. We have augmented the ordinal CMH tests to put them on the same basis as the NP ANOVA tests, but the big difference between the two methodologies is that NP ANOVA is much more generally applicable.

It is of interest that two of the elementary designs, the CRD and the RBD, are CMH designs and the traditional nonparametric tests are particular cases of ordinal CMH tests. It was something of a challenge to explore what might be considered the two next most elementary designs, the BIBD and the LSD, to see if parallel or related results could be found there. We were able to give new powerful rank tests for the LSD, but perhaps the most obvious outcome is that the NP ANOVA methodology applies seamlessly here, while CMH does not.

We hope that data analysts will feel motivated to apply the results outlined in this text and that researchers will consider our results and further develop them.

Bibliography

Agresti, A. (2003). *Categorical Data Analysis.* Hoboken, NJ: John Wiley & Sons.

Lancaster, H. O. (1949). The derivation and partition of $\chi 2$ in certain discrete distributions. *Biometrika*, 36(1-2):117–129.

Mantel, N. and Haenszel, W. (1959). Statistical aspects of the analysis of data from retrospective studies of disease. *Journal of the National Cancer Institute*, 22(4):719–748.

Appendix A

Appendix

These appendices contain material incidental to the main text. Specifically there is mathematical detail about the Moore–Penrose generalised inverse and Kronecker products.

A.1 Kronecker Products and Direct Sums

A Kronecker product is the same as a direct product. However, a direct sum and a Kronecker sum are different. Kronecker sums aren't relevant here.

The *direct sum* of two square matrices A and B, of sizes $m \times m$ and $n \times n$, respectively, is the square matrix of size $(m + n) \times (m + n)$, given by

$$A \oplus B = \begin{pmatrix} A & 0 \\ 0 & B \end{pmatrix}.$$

It is routine to show that

$$A \oplus (B \oplus C) = (A \oplus B) \oplus C;$$
$$(A \oplus B) + (C \oplus D) = (A + C) \oplus (B + D);$$
$$(A \oplus B)(C \oplus D) = AC \oplus BD;$$
$$(A \oplus B)^T = A^T \oplus B^T; \tag{A.1}$$
$$I_m \oplus I_n = I_{m+n};$$
$$(A \oplus B)^{-1} = A^{-1} \oplus B^{-1}; \text{ and}$$
$$\det(A \oplus B) = \det(A)\det(B).$$

For example, $(A \oplus B)\left(A^{-1} \oplus B^{-1}\right) = AA^{-1} \oplus BB^{-1} = I_{m+n}$.

An Introduction to Cochran–Mantel–Haenszel Testing and Nonparametric ANOVA, First Edition. J.C.W. Rayner and G. C. Livingston Jr.
© 2023 John Wiley & Sons Ltd. Published 2023 by John Wiley & Sons Ltd.

The direct or Kronecker product of two square matrices A and B, of sizes $m \times m$ and $n \times n$ respectively, is the square matrix of size $mn \times mn$, given by

$$A \otimes B = \begin{pmatrix} a_{11}B & a_{12}B & \cdots & a_{1m}B \\ a_{21}B & a_{22}B & \cdots & a_{2m}B \\ \vdots & \vdots & \ddots & \vdots \\ a_{m1}B & a_{m2}B & \cdots & a_{mm}B \end{pmatrix}$$

If $A = \{a_{ij}\}$ and $B = \{b_{rs}\}$, then $a_{ij}b_{rs} = (A \otimes B)_{(i-1)n+r,(j-1)n+s}$.
It may be shown that for conforming matrices

$$A \otimes (B \otimes C) = (A \otimes B) \otimes C;$$
$$(A \otimes B)(C \otimes D) = AC \otimes BD, \text{ so that, for example,}$$
$$(A \otimes B)(C \otimes D)(E \otimes F) = ACE \otimes CDF;$$
$$(A \otimes B)^T = A^T \otimes B^T; \tag{A.2}$$
$$I_m \otimes I_n = I_{mn};$$
$$(A \otimes B)^{-1} = A^{-1} \otimes B^{-1};$$
$$\text{trace}(A \otimes B) = \text{trace}(A)\, \text{trace}(B); \text{ and}$$
$$\det(A \otimes B) = \det(A)^m \det(B)^n.$$

It is useful to define the direct product of two vectors so that the elements of the product are arranged in dictionary or lexicographic order. Thus if $x = (x_1, \ldots, x_m)^T$ and if $y = (y_1, \ldots, y_n)^T$, then define

$$(x \otimes y)^T = (x_1 y_1, \ldots, x_1 y_n, \ldots, x_m y_1, \ldots, x_m y_n)^T.$$

Now suppose that A has eigenvalue λ with corresponding eigenvector x, and B has eigenvalue μ with corresponding eigenvector y. This means for $x \neq 0$, $Ax = \lambda x$ and for $y \neq 0$, $By = \mu y$. It follows that $A \otimes B$ has eigenvalue $\lambda\mu$ with corresponding eigenvector $x \otimes y$, since $x \otimes y \neq 0$ and

$$(A \otimes B)(x \otimes y) = \begin{pmatrix} a_{11}B & a_{12}B & \cdots & a_{1m}B \\ a_{21}B & a_{22}B & \cdots & a_{2m}B \\ \vdots & \vdots & \ddots & \vdots \\ a_{m1}B & a_{m2}B & \cdots & a_{mm}B \end{pmatrix} \begin{pmatrix} x_1 y \\ x_2 y \\ \vdots \\ x_m y \end{pmatrix}$$

$$= \begin{pmatrix} a_{11}x_1By & a_{12}x_2By & \cdots & a_{1m}x_mBy \\ a_{21}x_1By & a_{22}x_2By & \cdots & a_{2m}x_mBy \\ \vdots & \vdots & \ddots & \vdots \\ a_{m1}x_1By & a_{m2}x_2By & \cdots & a_{mm}x_mBy \end{pmatrix} \tag{A.3}$$

$$= \begin{pmatrix} a_{11}x_1 & a_{12}x_2 & \cdots & a_{1m}x_m \\ a_{21}x_1 & a_{22}x_2 & \cdots & a_{2m}x_m \\ \vdots & \vdots & \ddots & \vdots \\ a_{m1}x_1 & a_{m2}x_2 & \cdots & a_{mm}x_m \end{pmatrix} \otimes By$$

$$= Ax \otimes By = \lambda x \otimes \mu y = (\lambda \mu)(x \otimes y).$$

A.2 The Moore–Penrose Generalised Inverse

One of several pseudo-inverses or generalised inverses is the Moore–Penrose inverse. The unique Moore–Penrose inverse B^+ of a real symmetric matrix B satisfies

$$\begin{aligned} B^+BB^+ &= B^+, \\ BB^+B &= B, \\ \left(B^+B\right)^T &= B^+B \text{ and} \\ \left(BB^+\right)^T &= BB^+. \end{aligned} \tag{A.4}$$

It is routine to show that

- If $\Lambda = \text{diag}\left(\lambda_1, \ldots, \lambda_r, 0, \ldots, 0\right)$ then $\Lambda^+ = \text{diag}\left(\lambda_1^{-1}, \ldots, \lambda_r^{-1}, 0, \ldots, 0\right)$.
- If H is orthogonal then $H^+ = H^T$.
- If A is idempotent then $A^+ = A$.
- If the subsequent matrix products are defined then $(BC)^+ = C^+B^+$ and $(ABC)^+ = C^+B^+A^+$.

It is well-known that if X is $N_p(0, \Sigma)$ with $\text{rank}(\Sigma) = r < p$ then $X^T\Sigma^+X$ has the distribution χ_r^2 where Σ^+ is a pseudo-inverse of Σ.

Subject Index

An Introduction to Cochran–Mantel–Haenszel Testing and Nonparametric ANOVA,
First Edition. J.C.W. Rayner and G. C. Livingston Jr.
© 2023 John Wiley & Sons Ltd. Published 2023 by John Wiley & Sons Ltd.

References Index

a

Agresti (2003)   6, 206, 209
Akritas (1990)   162

b

Best and Rayner (2017a)   119
Best and Rayner (2017b)   118
Best and Rayner (2011)   151
Best and Rayner (2014)   41, 79, 80
Brockhoff, Best and Rayner (2004)
    80

c

Cochran (1950)   117
Cochran (1954)   9
Conover (1998)   76
Conover and Imam (1981)   4

d

Davis (2002)   13, 23, 36

e

Emerson (1968)   100

g

Gibbons and Chakraborti (2021)   32

h

Hodges and Lehmann (1962)   159

i

Irwin (1949)   10

k

Kuehl (2000)   75, 151
Kuritz, Landis and Koch (1988)   13
Kutner, Nachtsheim, Neter and Li
    (2005)   78

l

Lancaster (1949)   10, 214
Landis, Cooper, Kennedy and Koch
    (1979)   13, 45, 69
Landis, Cooper, Kennedy, and Koch
    (1979)   9
Landis, Heyman and Koch (1978)   9,
    10, 13, 16, 19, 20, 45

m

Manly (2007)   107
Mantel and Haenszel (1959)   9, 11,
    214
Maxwell (1970)   114

o

O'Mahony (1986)   23

p

Pearce (1960)   5, 139

An Introduction to Cochran–Mantel–Haenszel Testing and Nonparametric ANOVA,
First Edition. J.C.W. Rayner and G. C. Livingston Jr.
© 2023 John Wiley & Sons Ltd. Published 2023 by John Wiley & Sons Ltd.

Data Index

An Introduction to Cochran–Mantel–Haenszel Testing and Nonparametric ANOVA,
First Edition. J.C.W. Rayner and G. C. Livingston Jr.
© 2023 John Wiley & Sons Ltd. Published 2023 by John Wiley & Sons Ltd.